EUCLID'S BOOK ON DIVISIONS OF FIGURES

EUCLID'S BOOK ON DIVISIONS OF FIGURES

RAYMOND CLARE ARCHIBALD, Ph.D.
ASSISTANT PROFESSOR OF MATHEMATICS IN BROWN UNIVERSITY, PROVIDENCE, RHODE ISLAND

HAWK PRESS

Published by

Hawk Press
4836/24, Ansari Road, Daryaganj
New Delhi – 110 002
Phones : 9643330713, 011-23278618, 011-35676207
E-mail: thehawkpress@gmail.com
www.thehawkpress.com

ISBN: 978-93-93971-30-2

TO

MY OLD TEACHER AND FRIEND

ALFRED DEANE SMITH

PROFESSOR OF GREEK AND LATIN

AT MOUNT ALLISON UNIVERSITY

FOR FORTY-FOUR YEARS

SCHOLAR OF GREAT ATTAINMENTS

THE WONDER OF ALL WHO KNOW HIM

THESE PAGES ARE AFFECTIONATELY DEDICATED

INTRODUCTORY

Euclid, famed founder of the Alexandrian School of Mathematics, was the author of not less than nine works. Approximately complete texts, all carefully edited, of four of these, (1) the *Elements*, (2) the *Data*, (3) the *Optics*, (4) the *Phenomena*, are now our possession. In the case of (5) the *Pseudaria*, (6) the *Surface-Loci*, (7) the *Conics*, our fragmentary knowledge, derived wholly from Greek sources, makes conjecture as to their content of the vaguest nature. On (8) the *Porisms*, Pappus gives extended comment. As to (9), the book *On Divisions* (*of figures*), Proclus alone among Greeks makes explanatory reference. But in an Arabian MS., translated by Woepcke into French over sixty years ago, we have not only the enunciations of all of the propositions but also the proofs of four of them.

Whilst elaborate restorations of the *Porisms* by Simson and Chasles have been published, no previous attempt has been made (the pamphlet of Ofterdinger is not forgotten) to restore the proofs of the book *On Divisions* (*of figures*). And, except for a short sketch in Heath's monumental edition of Euclid's *Elements*, nothing but passing mention of Euclid's book *On Divisions* has appeared in English.

In this little volume I have attempted:

(1) to give, with necessary commentary, a restoration of Euclid's work based on the Woepcke text and on a thirteenth century geometry of Leonardo Pisano.

(2) to take due account of the various questions which arise in connection with (*a*) certain MSS. of "Muhammed Bagdedinus," (*b*) the Dee-Commandinus book on divisions of figures.

(3) to indicate the writers prior to 1500 who have dealt with propositions of Euclid's work.

(4) to make a selection from the very extensive bibliography of the subject during the past 400 years.

In the historical survey the MSS. of "Muhammed Bagdedinus" play an important rôle, and many recent historians, for example Heiberg, Cantor, Hankel, Loria, Suter, and Steinschneider, have contributed to the discussion. As it is necessary for me to correct errors, major and minor, of all of these writers, considerable detail has to be given in the first part of the volume; the brief second part treats of writers on divisions before 1500; the third part contains the restoration proper, with its thirty-six propositions. The Appendix deals with literature since 1500.

A score of the propositions are more or less familiar as isolated problems of modern English texts, and are also to be found in many recent English,

German and French books and periodicals. But any approximately accurate restoration of the work as a whole, in Euclidean manner, can hardly fail of appeal to anyone interested in elementary geometry or in Greek mathematics of twenty-two centuries ago.

In the spelling of Arabian names, I have followed Suter.

It is a pleasure to have to acknowledge indebtedness to the two foremost living authorities on Greek Mathematics. I refer to Professor J. L. Heiberg of the University of Copenhagen and to Sir Thomas L. Heath of London. Professor Heiberg most kindly sent me the proof pages of the forthcoming concluding volume of Euclid's *Opera Omnia*, which contained the references to Euclid's book *On Divisions of Figures.* To Sir Thomas my debt is great. On nearly every page that follows there is evidence of the influence of his publications; moreover, he has read this little book in proof and set me right at several points, more especially in connection with discussions in Note 113 and Paragraph 50.

R. C. A.

BROWN UNIVERSITY,
June, 1915.

CONTENTS

I.

Proclus, and Euclid's book On Divisions.

1. Last in a list of Euclid's works "full of admirable diligence and skilful consideration," Proclus mentions, without comment, *περὶ διαιρέσεων βιβλίον* [1]. But a little later[2] in speaking of the conception or definition of *figure* and of the divisibility of a figure into others differing from it in kind, Proclus adds: "For the circle is divisible into parts unlike in definition or notion, and so is each of the rectilineal figures; this is in fact the business of the writer of the Elements in his Divisions, where he divides given figures, in one case into like figures, and in another into unlike[3]."

De Divisionibus by Muhammed Bagdedinus and the Dee MS.

2. This is all we have from Greek sources, but the discovery of an Arabian translation of the treatise supplies the deficiency. In histories of Euclid's works (for example those by Hankel[4], Heiberg[5], Favaro[6], Loria[7], Cantor[8], Hultsch[9], Heath[3]) prominence is given to a treatise *De Divisionibus*, by one "Muhammed Bagdedinus." Of this in 1563[10] a copy (in Latin) was given by John Dee to

---

[1] *Procli Diadochi in primum Euclidis elementorum librum commentarii* ex rec. G. Friedlein, Leipzig, 1873, p. 69. Reference to this work will be made by "Proclus."

[2] Proclus[1], p. 144.

[3] In this translation I have followed T. L. Heath, *The Thirteen Books of Euclid's Elements*, 1, Cambridge, 1908, p. 8. To Heath's account (pp. 8–10) of Euclid's book *On Divisions* I shall refer by "Heath."

"Like" and "unlike" in the above quotation mean, not "similar" and "dissimilar" in the technical sense, but "like" or "unlike *in definition* or *notion*": thus to divide a triangle into triangles would be to divide it into "like" figures, to divide a triangle into a triangle and a quadrilateral would be to divide it into "unlike" figures. (Heath.)

[4] H. Hankel, *Zur Geschichte der Mathematik*, Leipzig, 1874, p. 234.

[5] J. L. Heiberg, *Litterargeschichtliche Studien über Euklid*, Leipzig, 1882, pp. 13–16, 36–38. Reference to this work will be made by "Heiberg."

[6] E. A. Favaro. "Preliminari ad una Restituzione del libro di Euclide sulla divisione delle figure piane," *Atti del reale Istituto Veneto di Scienze, Lettere ed Arti*, I$_6$, 1883, pp. 393–6. "Notizie storico-critiche sulla Divisione delee Aree" (Presentata li 28 gennaio, 1883), *Memorie del reale Istituto Veneto di Scienze, Lettere ed Arti*, XXII, 129–154. This is by far the most elaborate consideration of the subject up to the present. Reference to it will be made by "Favaro."

[7] G. Loria, "Le Scienze esatte nell' antica Grecia, Libro II, Il periodo aureo della geometria Greca." *Memorie della regia Accademia di Scienze, Lettere ed Arti in Modena*, XI$_2$, 1895, pp. 68–70, 220–221. *Le Scienze esatte nell' antica Grecia*, Seconda edizione. Milano, 1914, pp. 250–252, 426–427.

[8] M. Cantor, *Vorlesungen über Geschichte der Mathematik*, I$_3$, 1907, pp. 287–8; II$_2$, 1900, p. 555.

[9] F. Hultsch, Article "Eukleides" in Pauly-Wissowa's *Real-Encyclopädie der Class. Altertumswissenschaften*, VI, Stuttgart, 1909, especially Cols. 1040–41.

[10] When Dee was in Italy visiting Commandinus at Urbino.

Commandinus who published it in Dee's name and his own in 1570[11]. Recent writers whose publications appeared before 1905 have generally supposed that Dee had somewhere discovered an Arabian original of Muhammed's work and had given a Latin translation to Commandinus. Nothing contrary to this is indeed explicitly stated by Steinschneider when he writes in 1905[12], "Machomet Bagdadinus (=aus Bagdad) heisst in einem alten MS. Cotton (jetzt im Brit. Mus.) der Verfasser von: de Superficierum divisione (22 Lehrsätze); Jo. Dee aus London entdeckte es und übergab es T. Commandino...." For this suggestion as to the place where Dee found the MS. Steinschneider gives no authority. He does, however, give a reference to Wenrich[13], who in turn refers to a list of the printed books ("Impressi") of John Dee, in a life of Dee by Thomas Smith[14] (1638–1710). We here find as the third in the list, "Epistola ad eximium Ducis Urbini Mathematicum, Fredericum Commandinum, praefixa libello Machometi

---

[11] *De superficierum divisionibus liber Machometo Bagdedino ascriptus nunc primum Joannis Dee Londinensis & Federici Commandini Urbinatis opera in lucem editus.* Federici Commandini de eadem re libellus. Pisauri, MDLXX. In the same year appeared an Italian translation: *Libro del modo di dividere le superficie attribuito a Machometo Bagdedino. Mandato in luce la prima volta da M. G. Dee... e da M. F.... Commandino... Tradotti dal Latino in volgare da F. Viani de' Malatesti,...* In Pesaro, del MDLXX...4 unnumbered leaves and 44 numbered on one side.

An English translation from the Latin, with the following title-page, was published in the next century: *A Book of the Divisions of Superficies: ascribed to Machomet Bagdedine. Now put forth, by the pains of John Dee of London, and Frederic Commandine of Urbin. As also a little Book of Frederic Commandine, concerning the same matter. London Printed by R. & W. Leybourn,* 1660. Although this work has a separate title page and the above date, it occupies the last fifty pages (601–650) of a work dated a year later: *Euclid's Elements of Geometry in XV Books... to which is added a Treatise of Regular Solids by Campane and Flussas likewise Euclid's Data and Marinus Preface thereunto annexed. Also a Treatise of the Divisions of Superficies ascribed to Machomet Bagdedine, but published by Commandine, at the request of John Dee of London; whose Preface to the said Treatise declares it to be the Worke of Euclide, the Author of the Elements. Published by the care and Industry of John Leeke and George Serle, Students in the Mathematics.* London... MDCLXI.

A reprint of simply that portion of the Latin edition which is the text of Muhammed's work appeared in: *ΕΥΚΛΕΙΔΟΥ ΤΑ ΣΩΖΟΜΕΝΑ Euclidis quae supersunt omnia. Ex rescensione Davidis Gregorii...* Oxoniae... MDCCIII. Pp. 665–684: *ΕΥΚΛΕΙΔΟΥ ΩΣ ΟΙΟΝΤΑΙ ΤΙΝΕΣ, ΠΕΡΙ ΔΙΑΙΡΕΣΕΩΝ ΒΙΒΛΟΣ* Euclidis, ut quidam arbitrantur, de divisionibus liber—vel ut alii volunt, Machometi Bagdedini liber de divisionibus superficierum."

[12] M. STEINSCHNEIDER, "Die Europäischen Übersetzungen aus dem Arabischen bis Mitte des 17. Jahrhunderts." *Sitzungsberichte der Akademie der Wissenschaften in Wien* (Philog.-histor. Klasse) CLI, Jan. 1905, Wien, 1906. Concerning "171. Muhammed" *cf.* pp. 41–2. Reference to this paper will be made by "Steinschneider."

[13] J. G. WENRICH, *De auctorum Graecorum versionibus.* Lipsiae, MDCCCXLII, p. 184.

[14] T. SMITH, *Vitae quorundam eruditissimorum et illustrium virorum...* Londini... MDCCVII, p. 56. It was only the first 55 pages of this "Vita Joannis Dee, Mathematici Angli," which were translated into English by W. A. Ayton, London, 1908.

Bagdedini de superficierum divisionibus... *Pisauri*, 1570. Exstat MS. in Bibliotheca Cottoniana sub Tiberio B IX."

Then come the following somewhat mysterious sentences which I give in translation[15]: "After the preface Lord Ussher [1581–1656], Archbishop of Armagh, has these lines: It is to be noted that the author uses Euclid's Elements translated into the Arabic tongue, which Campanus afterwards turned into Latin. Euclid therefore seems to have been the author of the Propositions [of *De Divisionibus*] though not of the demonstrations, which contain references to an Arabic edition of the Elements, and which are due to Machometus of Bagded or Babylon." This quotation from Smith is reproduced, with various changes in punctuation and typography, by Kästner[16]. Consideration of the latter part of it I shall postpone to a later article (5).

3. Following up the suggestion of Steinschneider, Suter pointed out[17], without reference to Smith[14] or Kästner[16], that in Smith's catalogue of the Cottonian Library there was an entry[18] under "Tiberius[19] B IX, 6": "Liber Divisionum Mahumeti Bag-dadini." As this MS. was undoubtedly in Latin and as Cottonian MSS. are now in the British Museum, Suter inferred that Dee simply made a copy of the above mentioned MS. and that this MS. was now in the British Museum. With his wonted carefulness of statement, Heath does not commit himself to these views although he admits their probable accuracy.

4. As a final settlement of the question, I propose to show that Steinschneider and Suter, and hence also many earlier writers, have not considered all facts available. Some of their conclusions are therefore untenable. In particular:

(1) In or before 1563 Dee did *not* make a copy of any Cottonian MS.;

(2) The above mentioned MS. (Tiberius, B. IX, 6) was never, in its entirety, in the British Museum;

---

15 "Post praefationem haec habet D. *Usserius* Archiepiscopus Armachanus. *Notandum est autem, Auctorem hunc* Euclide *usum in Arabicam linguam converso, quem postea* Campanus *Latinum fecit. Auctor igitur propositionum videtur fuisse* Euclides: *demonstrationum, in quibus* Euclides *in Arabico codice citatur*, Machometus Bagded *sive* Babylonius."

It has been stated that Campanus (13. cent.) did not translate Euclid's Elements into Latin, but that the work published as his (Venice, 1482—the first printed edition of the *Elements*) was the translation made about 1120 by the English monk Athelhard of Bath. *Cf.* HEATH, *Thirteen Books of Euclid's Elements*, I, 78, 93–96.

16 A. G. KÄSTNER, *Geschichte der Mathematik*... Erster Band... Göttingen, 1796, pp. 272–3. See also "Zweyter" Band, 1797, pp. 46–47.

17 H. SUTER, "Zu dem Buche 'De Superficierum divisionibus' des Muhammed Bagdedinus." *Bibliotheca Mathematica*, VI$_3$, 321–2, 1905.

18 T. SMITH, *Catalogus Librorum Manuscriptorum Bibliothecae Cottonianae*... Oxonii,... MDCXCVI, p. 24.

19 The original Cottonian library was contained in 14 presses, above each of which was a bust; 12 of these busts were of Roman Emperors. Hence the classification of the MSS. in the catalogue.

(3) The inference by Suter that this MS. was probably the Latin translation of the tract from the Arabic, made by Gherard of Cremona (1114–1187)—among the lists of whose numerous translations a "liber divisionum" occurs—*should be accepted with great reserve*;

(4) The MS. which Dee used can be stated with absolute certainty and this MS. did not, in all probability, afterwards become a Cottonian MS.

(1) Sir Robert Bruce Cotton, the founder of the Cottonian Library, was born in 1571. The Cottonian Library was not, therefore, in existence in 1563 and Dee could not then have copied a Cottonian MS.

(2) The Cottonian Library passed into the care of the nation shortly after 1700. In 1731 about 200 of the MSS. were damaged or destroyed by fire. As a result of the parliamentary inquiry Casley reported[20] on the MSS. destroyed or injured. Concerning Tiberius IX, he wrote, "This volume burnt to a crust." He gives the title of each tract and the folios occupied by each in the volume. "Liber Divisionum Mahumeti Bag-dadini" occupied folios 254–258. When the British Museum was opened in *1753*, what was left of the Cottonian Library was immediately placed there. Although portions of all of the leaves of our tract are now to be seen in the British Museum, practically none of the writing is decipherable.

(3) Planta's catalogue[21] has the following note concerning Tiberius IX: "A volume on parchment, which once consisted of 272 leaves, written about the XIV. century [not the XII. century, when Gherard of Cremona flourished], containing eight tracts, the principal of which was a 'Register of William Cratfield, abbot

---

[20] D. CASLEY, p. 15 ff. of *A Report from the Committee appointed to view the Cottonian Library... Published by order of the House of Commons.* London, MDCCXXXII (British Museum MSS. 24932). *Cf.* also the page opposite that numbered 120 in *A Catalogue of the Manuscripts in the Cottonian Library... with an Appendix containing an account of the damage sustained by the Fire in 1731; by S. Hooper...* London:... MDCCLXXVII.

[21] J. PLANTA, *A Catalogue of the Manuscripts in the Cottonian Library deposited in the British Museum. Printed by command of his Majesty King George III...* 1802.

In the British Museum there are three MS. catalogues of the Cottonian Library:

(1) *Harleian MS.* 6018, a catalogue made in 1621. At the end are memoranda of loaned books. On a sheet of paper bearing date Novem. 23, 1638, Tiberius B IX is listed (folio 187) with its art. 4: "liber divisione Machumeti Bagdedini." The paper is torn so that the name of the person to whom the work was loaned is missing. The volume is not mentioned in the main catalogue.

(2) *MS. No.* 36789, made after Sir Robert Cotton's death in 1631 and before 1638 (*cf. Catalogue of Additions to the MSS. in British Museum, 1900–1905...* London, 1907, pp. 226–227), contains, apparently, no reference to "Muhammed."

(3) *MS. No.* 36682 A, of uncertain date but earlier than 1654 (*Catalogue of Additions... l.c.* pp. 188–189). On folio 78 *verso* we find Tiberius B IX, Art. 4: "Liber divisione Machumeti Bagdedini."

A "Muhammed" MS. was therefore in the Cottonian Library in 1638.

The anonymously printed (1840?) "Index to articles printed from the Cotton MSS., & where they may be found" which may be seen in the British Museum, only gives references to the MSS. in "Julius."

of St Edmund' " [d. 1415]. Tracts 3, 4, 5 were on music.

(4) On "*A*° 1583, 6 Sept." Dee made a catalogue of the MSS. which he owned. This catalogue, which is in the Library of Trinity College, Cambridge[22], has been published[23] under the editorship of J. O. Halliwell. The 95th item described is a folio parchment volume containing 24 tracts on mathematics and astronomy. The 17th tract is entitled "Machumeti Bagdedini liber divisionum." As the contents of this volume are entirely different from those of Tiberius IX described above, in (3), it seems probable that there were two copies of "Muhammed's" tract, while the MS. which Dee used for the 1570 publication was undoubtedly his own, as we shall presently see. If the two copies be granted, there is no evidence against the Dee copy having been that made by Gherard of Cremona.

5. There is the not remote possibility that the Dee MS. was destroyed soon after it was catalogued. For in the same month that the above catalogue was prepared, Dee left his home at Mortlake, Surrey, for a lengthy trip in Europe. Immediately after his departure "the mob, who execrated him as a magician, broke into his house and destroyed a great part of his furniture and books[24]. . . " many of which "were the written bookes[25]." Now the Dee catalogue of his MSS. (MS. O. iv. 20), in Trinity College Library, has numerous annotations[26] in Dee's handwriting. They indicate just what works were (1) destroyed or stolen ("Fr.")[27] and (2) left("T.")[28] after the raid. Opposite the titles of the tracts in the volume including the tract "liber divisionum," "Fr." is written, and opposite the title "Machumeti Bagdedini liber divisionum" is the following note: "Curavi imprimi Urbini in Italia per Federicum Commandinum exemplari descripto ex vetusto isto monumento(?) per me ipsum." Hence, as stated above, it is now definitely known (1) that the MS. which Dee used was his own, and (2) that some 20 years after he made a copy, the MS. was stolen and probably destroyed[29].

On the other hand we have the apparently contradictory evidence in the passage quoted above (Art. 2) from the life of Dee by Smith[14] who was also the compiler of the Catalogue of the Cottonian Library. Smith was librarian when he wrote both of these works, so that any definite statement which he

---

[22] A transcription of the Trinity College copy, by Ashmole, is in MS. Ashm. 1142. Another autograph copy is in the British Museum: Harleian MS. 1879.

[23] *Camden Society Publications*, XIX, London, M.DCCC.XLII.

[24] *Dictionary of National Biography*, Article, "Dee, John."

[25] "The compendious rehearsall of John Dee his dutifull declaration A. 1592" printed in *Chetham Miscellanies*, vol. I, Manchester, 1851, p. 27.

[26] Although Halliwell professed to publish the Trinity MS., he makes not the slightest reference to these annotations.

[27] "Fr." is no doubt an abbreviation for *Furatum*.

[28] "T.", according to Ainsworth (*Latin Dictionary*), was put after the name of a soldier to indicate that he had survived (*superstes*). Whence this abbreviation?

[29] The view concerning the theft or destruction of the MS. is borne out by the fact that in a catalogue of Dee's Library (British Museum MS. 35213) made early in the seventeenth century (*Catalogue of Additions and Manuscripts*. . . 1901, p. 211), Machumeti Bagdedini is not mentioned.

makes concerning the library long in his charge is not likely to be successfully challenged. Smith does not however say that Dee's "Muhammed" MS. was in the Cottonian Library, and if he knew that such was the case we should certainly expect some note to that effect in the catalogue[18]; for in three other places in his catalogue (Vespasian B X, A II$_{13}$, Galba E VIII), Dee's original ownership of MSS. which finally came to the Cottonian Library is carefully remarked. Smith does declare, however, that the Cottonian MS. bore, "after the preface," certain notes (which I have quoted above) by Archbishop Ussher (1581–1656). Now it is not a little curious that these notes by Ussher, who was not born till after the Dee book was printed, should be practically identical with notes in the printed work, just after Dee's letter to Commandinus (Art. 3). For the sake of comparison I quote the notes in question[30]; "To the Reader.—I am here to advertise thee (kinde Reader) that this author which we present to thee, made use of Euclid translated into the Arabick Tongue, whom afterwards Campanus made to speake Latine. This I thought fit to tell thee, that so in searching or examining the Propositions which are cited by him, thou mightest not sometime or other trouble thy selfe in vain, Farewell."

The Dee MS. as published did not have any preface. We can therefore only assume that Ussher wrote in a MS. which *did* have a preface the few lines which he may have seen in Dee's printed book.

6. Other suggestions which have been made concerning "Muhammed's" tract should be considered. Steinschneider asks, "Ob identisch de Curvis superficiebus, von einem Muhammed, MS. Brit. Mus. Harl. 623^6 (I, 191)[31]?" I have examined this MS. and found that it has nothing to do with the subject matter of the Dee tract.

But again, Favaro states[32]: "Probabilmente il manoscritto del quale si servì il Dee è lo stesso indicato dall'Heilbronner[33] come esistente nella Biblioteca Bodleiana di Oxford." Under date "6. 3. 1912" Dr A. Cowley, assistant librarian in the Bodleian, wrote me as follows: "We do not possess a copy of Heilbronner's Hist. Math. Univ. In the old catalogue of MSS. which he would have used, the work you mention is included—but is really a printed book and is only included in the catalogue of MSS. because it contains some manuscript notes—

"Its shelf-mark is Savile T 20.

"It has 76 pages in excellent condition. The title page has: De Superficierum | divisionibus liber | Machometo Bagdedino | ascriptus | nunc primum Joannis Dee | ... | opera in lucem editus | ... Pisauri MDLXX.

---

30 This quotation from the Leeke-Serle Euclid[11] is an exact translation of the original.

31 This should be 625^6 (I, 391).

32 Favaro, p. 140. *Cf.* Heiberg, p. 14. This suggestion doubtless originated with Ofterdinger[38], p. [1].

33 J. C. HEILBRONNER, *Historia matheseos Universae*... Lipsiae, MDCCXLII, p. 620: ("Manuscripta mathematica in Bibliotheca Bodlejana") "34 Mohammedis Bagdadeni liber de superficierum divisionibus, cum Notis H. S."

"The MS. notes are by Savile, from whom we got the collection to which this volume belongs."

The notes were incorporated into the Gregory edition[11] of the Dee tract. Here and elsewhere[34] Savile objected to attributing the tract to Euclid as author[35]. His arguments are summed up, for the most part, in the conclusions of Heiberg followed by Heath: "the Arabic original could not have been a direct translation from Euclid, and probably was not even a direct adaptation of it; it contains mistakes and unmathematical expressions, and moreover does not contain the propositions about the division of a circle alluded to by Proclus. Hence it can scarcely have contained more than a fragment of Euclid's work."

The Woepcke-Euclid MS.

7. On the other hand Woepcke found in a MS. (No. 952. 2 Arab. Suppl.) of the Bibliothèque nationale, Paris, a treatise in Arabic on the division of plane figures, which he translated, and published in 1851[36]. "It is expressly

---

34 H. SAVILE, *Praelectiones tresdecim in principium elementorum Evclidis, Oxonii habitae M.DC.XX.* Oxonii..., 1621, pp. 17–18.

35 Dee's statement of the case in his letter to Commandinus (Leeke-Serle Euclid[11], *cf.* note 30) is as follows: "As for the authors name, I would have you understand, that to the very old Copy from whence I writ it, the name of MACHOMET BAGDEDINE was put in ziphers or Characters, (as they call them) who whether he were that *Albategnus* whom *Copernicus* often cites as a very considerable Author in Astronomie; or that Machomet who is said to have been *Alkindus's* scholar, and is reported to have written somewhat of the art of Demonstration, I am not yet certain of: or rather that this may be deemed a Book of our *Euclide*, all whose Books were long since turned out of the Greeke into the Syriack and Arabick Tongues. Whereupon, It being found some time or other to want its Title with the *Arabians* or *Syrians*, was easily attributed by the transcribers to that most famous Mathematician among them, Machomet: which I am able to prove by many testimonies, to be often done in many Moniments of the Ancients; ...yea further, we could not yet perceive so great acuteness of any *Machomet* in the Mathematicks, from their moniments which we enjoy, as everywhere appears in these Problems. Moreover, that *Euclide* also himself wrote one Book περι διαιρέσεων that is to say, *of Divisions*, as may be evidenced from Proclus's Commentaries upon his first of *Elements*: and we know none other extant under this title, nor can we find any, which for excellencie of its treatment, may more rightfully or worthily be ascribed to *Euclid*. Finally, I remember that in a certain very ancient piece of Geometry, I have read a place cited out of this little Book in expresse words, even as from amost (*sic*) certain work of *Euclid*. Therefore we have thus briefly declared our opinions for the present, which we desire may carry with them so much weight, as they have truth in them.... But whatsoever that Book of *Euclid* was concerning Divisions, certainly this is such an one as may be both very profitable for the studies of many, and also bring much honour and renown to every most noble ancient Mathematician; for the most excellent acutenesse of the invention, and the most accurate discussing of all the Cases in each Probleme...."

36 F. WOEPCKE, "Notice sur des traductions Arabes de deux ouverages perdus d'Euclide" *Journal Asiatique*, Septembre–Octobre, 1851, XVIII_4, 217–247. Euclid's work *On the division (of plane figures)*: pp. 233–244. Reference to this paper will be made by "Woepcke." In *Euclidis opera omnia*, vol. 8, now in the press, there are "Fragmenta

attributed to Euclid in the MS. and corresponds to the description of it by Proclus. Generally speaking, the divisions are divisions into figures of the same kind as the original figures, e. g. of triangles into triangles; but there are also divisions into 'unlike' figures, e. g. that of a triangle by a straight line parallel to the base. The missing propositions about the division of a circle are also here: 'to divide into two equal parts a given figure bounded by an arc of a circle and two straight lines including a given angle' and 'to draw in a given circle two parallel straight lines cutting off a certain part of a circle.' Unfortunately the proofs are given of only four propositions (including the two last mentioned) out of 36, because the Arabian translator found them too easy and omitted them." That the omission is due to the translator and did not occur in the original is indicated in two ways, as Heiberg points out. Five auxiliary propositions (Woepcke 21, 22, 23, 24, 25) of which no use is made are introduced. Also Woepcke 5 is: "...and we divide the triangle by a construction analogous to the preceding construction"; but no such construction is given.

The four proofs that are given are elegant and depend only on the propositions (or easy deductions from them) of the *Elements*, while Woepcke 18 has the true Greek ring: "to apply to a straight line a rectangle equal to the rectangle contained by AB, AC and *deficient by a square*."

8. To no proposition in the Dee MS. is there word for word correspondence with the propositions of Woepcke but in content there are several cases of likeness. Thus, Heiberg continues,

Dee 3 = Woepcke 30 (a special case is Woepcke 1);
Dee 7 = Woepcke 34 (a special case is Woepcke 14);
Dee 9 = Woepcke 36 (a special case is Woepcke 16);
Dee 12 = Woepcke 32 (a special case is Woepcke 4).

Woepcke 3 is only a special case of Dee 2; Woepcke 6, 7, 8, 9 are easily solved by Dee 8. And it can hardly be chance that the proofs of exactly these propositions in Dee should be without fault. That the treatise published by Woepcke is no fragment but the complete work which was before the translator is expressly stated[37], "fin du traité." It is moreover a well ordered and compact whole. Hence we may safely conclude that Woepcke's is not only Euclid's own work but the whole of it, except for proofs of some propositions.

---

collegit et disposuit J. L. Heiberg," through whose great courtesy I have been enabled to see the proof-sheets. First among the fragments, on pages 227–235, are (1) the Proclus references to περι διαιρέσεων and (2) the Woepcke translation mentioned above. In the article on Euclid in the last edition of the *Encyclopaedia Britannica* no reference is made to this work or to the writings of Heiberg, Hultsch, Steinschneider and Suter.

[37] Woepcke, p. 244.

9. For the reason just stated the so-called *Wiederherstellung* of Euclid's work by Ofterdinger[38], based mainly on Dee, is decidedly misnamed. A more accurate description of this pamphlet would be, "A translation of the Dee tract with indications in notes of a certain correspondence with 15 of Woepcke's propositions, the whole concluding with a translation of the enunciations of 16 of the remaining 21 propositions of Woepcke not previously mentioned." Woepcke 30, 31, 34, 35, 36 are not even noticed by Ofterdinger. Hence the claim I made above ("Introductory") that the first real restoration of Euclid's work is now presented. Having introduced Woepcke's text as one part of the basis of this restoration, the other part demands the consideration of the

Practica Geometriae of Leonardo Pisano (Fibonaci).

10. It was in the year 1220 that Leonardo Pisano, who occupies such an important place in the history of mathematics of the thirteenth century[39], wrote his *Practica Geometriae*, and the MS. is now in the Vatican Library. Although it was known and used by other writers, nearly six and one half centuries elapsed before it was finally published by Prince Boncompagni[40]. Favaro was the first[6] to call attention to the importance of Section IIII[41] of the *Practica Geometriae* in connection with the history of Euclid's work. This section is wholly devoted to the enunciation and proof and numerical exemplification of propositions concerning the divisions of figures. Favaro reproduces the enunciations of the propositions and numbers them 1 to 57[42]. He points out that in both enunciation and proof Leonardo 3, 10, 51, 57 are identical with Woepcke 19, 20, 29, 28 respectively. But considerably more remains to be remarked.

11. No less than twenty-two of Woepcke's propositions are practically identical in statement with propositions in Leonardo; the solutions of eight more of Woepcke are either given or clearly indicated by Leonardo's methods, and all six of the remaining Woepcke propositions (which are auxiliary) are assumed as known in the proofs which Leonardo gives of propositions in Woepcke. Indeed, these two works have a remarkable similarity. Not only are practically all of the Woepcke propositions in Leonardo, but the proofs called for by the order of

---

[38] L. F. Ofterdinger, *Beiträge zur Wiederherstellung der Schrift des Euklides über der Theilung der Figuren*, Ulm, 1853.

[39] M. Cantor, *Vorlesungen über Geschichte der Mathematik*, II_2, 1900, pp. 3–53; "Practica Geometriae," pp. 35–40.

[40] *Scritti di Leonardo Pisano matematico del secolo decimoterzo pubblicati da Baldassarre Boncompagni.* Volume II (Leonardi Pisani Practica Geometriae ed opuscoli). Roma... 1862. Practica Geometriae, pp. 1–224.

[41] *Scritti di Leonardo Pisano*... II, pp. 110–148.

[42] These numbers I shall use in what follows. Favaro omits some auxiliary propositions and makes slips in connection with 28 and 40. Either 28 should have been more general in statement or another number should have been introduced. Similarly for 40. Compare Articles 33–34, 35.

the propositions and by the auxiliary propositions in Woepcke are, with a possible single exception[91], invariably the kind of proofs which Euclid might have given—no other propositions but those which had gone before or which were to be found in the *Elements* being required in the successive constructions.

Leonardo had a wide range of knowledge concerning Arabian mathematics and the mathematics of antiquity. His *Practica Geometriae* contains many references to Euclid's *Elements* and many uncredited extracts from this work[43]. Similar treatment is accorded works of other writers. But in the great elegance, finish and rigour of the whole, originality of treatment is not infrequently evident. If Gherard of Cremona made a translation of Euclid's book *On Divisions*, it is not at all impossible that this may have been used by Leonardo. At any rate the conclusion seems inevitable that he must have had access to some such MS. of Greek or Arabian origin.

Further evidence that Leonardo's work was of Greek-Arabic extraction can be found in the fact that, in connection with the 113 figures, of the section On Divisions, of Leonardo's work, the lettering in only 58 contains the letters c or f; that is, the Greek-Arabic succession $a, b, g, d, e, z \ldots$ is used almost as frequently as the Latin $a, b, c, d, e, f, g, \ldots$; elimination of Latin letters added to a Greek succession in a figure, for the purpose of numerical examples (in which the work abounds), makes the balance equal.

12. My method of restoration of Euclid's work has been as follows. Everything in Woepcke's text (together with his notes) has been translated literally, reproduced without change and enclosed by quotation marks. To all of Euclid's enunciations (unaccompanied by constructions) which corresponded to enunciations by Leonardo, I have reproduced Leonardo's constructions and proofs, with the same lettering of the figures[44], but occasional abbreviation in the form of statement; that is, the extended form of Euclid in Woepcke's text, which is also employed by Leonardo, has been sometimes abridged by modern notation or briefer statement. Occasionally some very obvious steps taken by Leonardo have been left out but all such places are clearly indicated by explanation in square brackets, [ ]. Unless stated to the contrary, and indicated by different type, no step is given in a construction or proof which is not contained in Leonardo. When there is no correspondence between Woepcke and Leonardo I have exercised care to reproduce Leonardo's methods in other propositions, as closely as possible. If, in a given proposition, the method is extremely obvious on account of what has gone before, I have sometimes given little more than an indication of the propositions containing the essence of the required construction and proof. In the case of the six auxiliary propositions, the proofs supplied seemed to be readily suggested by propositions in Euclid's *Elements*.

---

[43] For example, on pages 15–16, 38, 95, 100–1, 154.

[44] This is done in order to give indication of the possible origin of the construction in question (Art. 11).

13. Immediately after the enunciations of Euclid's problems follow the statements of the correspondence with Leonardo; if exact, a bracket encloses the number of the Leonardo proposition, according to Favaro's numbering, and the page and lines of Boncompagni's edition where Leonardo enunciates the same proposition.

The following is a comparative table of the Euclid and, in brackets, of the corresponding Leonardo problems: 1 (5); 2 (14); 3 (2, 1); 4 (23); 5 (33); 6 (16); 7 (20)[45]; 8 (27)[46]; 9 (30, 31)[47]; 10 (18); 11 (0); 12 (28)[42]; 13 (32)[47]; 14 (36); 15 (40); 16 (37); 17 (39); 18 (0); 19 (3); 20 (10); 21 (0); 22 (0); 23 (0); 24 (0); 25 (0); 26 (4); 27 (11); 28 (57); 29 (51)[45]; 30 (0); 31 (0); 32 (29); 33 (35); 34 (40)[42]; 35 (0); 36 (0).

Summary

It will be instructive, as a means of comparison, to set forth in synoptic fashion: (1) the Muhammed-Commandinus treatise; (2) the Euclid treatise; (3) Leonardo's work. In (1) and (2) I follow Woepcke closely[48].

14. *Synopsis of Muhammed's Treatise*—

I. In all the problems it is required to divide the proposed figure into two parts having a given ratio.

II. The figures divided are: the triangle (props. 1–6); the parallelogram (11); the trapezium[89] (8, 12, 13); the quadrilateral (7, 9, 14–16); the pentagon (17, 18, 22); a pentagon with two parallel sides (19), a pentagon of which a side is parallel to a diagonal (20).

III. The transversal required to be drawn:

A. passes through a given point and is situated:

1. at a vertex of the proposed figure (1, 7, 17);
2. on any side (2, 9, 18);
3. on one of the two parallel sides (8).

B. is parallel:

1. to a side (not parallel) (3, 13, 14, 22);
2. to the parallel sides (11, 12, 19);

[45] Leonardo considers the case of "one third" instead of Euclid's "a certain fraction," but in the case of 20 he concludes that in the same way the figure may be divided "into four or many equal parts." *Cf.* Article 28.

[46] Woepcke 8 may be considered as a part of Leonardo 27 or better as an unnumbered proposition following Leonardo 25.

[47] Leonardo's propositions 30–32 consider somewhat more general problems than Euclid's 9 and 13. *Cf.* Articles 30 and 34.

[48] Woepcke, pp. 245–246.

3. to a diagonal (15, 20);
4. to a perpendicular drawn from a vertex of the figure to the opposite side (4);
5. to a transversal which passes through a vertex of the figure (5);
6. to any transversal (6, 16).

IV. Prop. 10: Being given the segment AB and two lines which pass through the extremities of this segment and form with the line AB any angles, draw a line parallel to AB from one or the other side of AB and such as to produce a trapezium of given size.

Prop. 21. Auxiliary theorem regarding the pentagon.

15. *Commandinus's Treatise*—Appended to the first published edition of Muhammed's work was a short treatise[49] by Commandinus who said[50] of Muhammed: "for what things the author of the book hath at large comprehended in many problems, I have compendiously comprised and dispatched in two only." This statement repeated by Ofterdinger[51] and Favaro[52] is somewhat misleading.

The "two problems" of Commandinus are as follows:

"Problem I. To divide a right lined figure according to a proportion given, from a point given in any part of the ambitus or circuit thereof, whether the said point be taken in any angle or side of the figure."

"Problem II. To divide a right lined figure $GABC$, according to a proportion given, E to F, by a right line parallel to another given line D."

But the first problem is divided into 18 cases: 4 for the triangle, 6 for the quadrilateral, 4 for the pentagon, 2 for the hexagon and 2 for the heptagon; and the second problem, as Commandinus treats it, has 20 cases: 3 for the triangle, 7 for the quadrilateral, 4 for the pentagon, 4 for the hexagon, 2 for the heptagon.

16. *Synopsis of Euclid's Treatise*—

I. The proposed figure is divided:

1. into two equal parts (1, 3, 4, 6, 8, 10, 12, 14, 16, 19, 26, 28);
2. into several equal parts (2, 5, 7, 9, 11, 13, 15, 17, 29);
3. into two parts, in a given ratio (20, 27, 30, 32, 34, 36);
4. into several parts, in a given ratio (31, 33, 35, 36).

The construction 1 or 3 is always followed by the construction of 2 or 4, except in the propositions 3, 28, 29.

---

49 Commandinus[11], pp. 54–76.
50 Commandinus[11], p. [ii]; Leeke-Serle Euclid, p. 603.
51 Ofterdinger[38], p. 11, note.
52 Favaro[6], p. 139.

II. The figures divided are:

the triangle (1, 2, 3, 19, 20, 26, 27, 30, 31);

the parallelogram (6, 7, 10, 11);

the trapezium (4, 5, 8, 9, 12, 13, 32, 33);

the quadrilateral (14, 15, 16, 17, 34, 35, 36);

a figure bounded by an arc of a circle and two lines (28);

the circle (29).

III. It is required to draw a transversal:

A. passing through a point situated:

1. at a vertex of the figure (14, 15, 34, 35);
2. on any side (3, 6, 7, 16, 17, 36);
3. on one of two parallel sides (8, 9);
4. at the middle of the arc of the circle (28);
5. in the interior of the figure (19, 20);
6. outside the figure (10, 11, 26, 27);
7. in a certain part of the plane of the figure (12, 13)

B. parallel to the base of the proposed figure (1, 2, 4, 5, 30–33).

C. parallel to one another, the problem is indeterminate (29).

IV. Auxiliary propositions:

18. To apply to a given line a rectangle of given size and deficient by a square.

21, 22, when $a \cdot d \gtrless b \cdot c$, it follows that $a : b \gtrless c : d$;

23, 24, when $a : b > c : d$, it follows that

$$(a \mp b) : b > (c \mp d) : d;$$

25, when $a : b < c : d$, it follows that $(a - b) : b < (c - d) : d$.

In the synopsis of the last five propositions I have changed the original notation slightly.

17. *Analysis of Leonardo's Work.* I have not thought it necessary to introduce into this analysis the unnumbered propositions referred to above[42].

I. The proposed figure is divided:

1. into two equal parts (1–5, 15–18, 23–28, 36–38, 42–46, 53–55, 57);
2. into several equal parts (6, 7, 9, 13, 14, 19, 21, 33, 47–50, 56);

3. into two parts in a given ratio (8, 10–12, 20, 29–32, 34, 39, 40, 51, 52);
4. into several parts in a given ratio (22, 35, 41).

The construction 1 or 3 is always followed by the construction of 2 or 4 except in the propositions 42–46, 51, 54, 57.

II. The figures divided are:

the triangle (1–14);

the parallelogram (15–22);

the trapezium (23–35);

the quadrilateral (36–41);

the pentagon (42–43);

the hexagon (44);

the circle and semicircle (45–56);

a figure bounded by an arc of a circle and two lines (57).

III. (i) It is required to draw a transversal:

A. passing through a point situated:
   1. at a vertex of the figure (1, 6, 26, 31, 34, 36, 41–44);
   2. on a side not produced (2, 7, 8, 16, 20, 37, 39);
   3. at a vertex or a point in a side (40);
   4. on one of two parallel sides (24, 25, 27, 30);
   5. on the middle of the arc of the circle (53, 55, 57);
   6. on the circumference or outside of the circle (45);
   7. inside of the figure (3, 10, 15, 17, 46);
   8. outside of the figure (4, 11, 12, 18);
   9. either inside or outside of the figure (38);
   10. either inside or outside or on a side of the figure (32);
   11. in a certain part of the plane of the figure (28).

B. parallel to the base of the proposed figure (5, 14, 19, 21–23, 29, 33, 35, 54);

C. parallel to a diameter of the circle (49, 50).

(ii) It is required to draw more than one transversal (*a*) through one point (9, 47, 48, 56); (*b*) through two points (13); (*c*) parallel to one another, the problem is indeterminate (51).

(iii) It is required to draw a circle (52).

IV. Auxiliary Propositions:

Although not explicitly stated or proved, Leonardo makes use of four out of six of Euclid's auxiliary propositions[113]. On the other hand he proves two other propositions which Favaro does not number: (1) Triangles with one angle of the one equal to one angle of the other, are to one another as the rectangle formed by the sides about the one angle is to that formed by the sides about the equal angle in the other; (2) the medians of a triangle meet in a point and trisect one another.

II.

18. *Abraham Savasorda, Jordanus Nemorarius, Luca Paciuolo.*—In earlier articles (10, 11) incidental reference was made to Leonardo's general indebtedness to previous writers in preparing his *Practica Geometriae*, and also to the debt which later writers owe to Leonardo. Among the former, perhaps mention should be made of Abraham bar Chijja ha Nasi[53] of Savasorda and his *Liber embadorum* known through the Latin translation of Plato of Tivoli. Abraham was a learned Jew of Barcelona who probably employed Plato of Tivoli to make the translation of his work from the Hebrew. This translation, completed in 1116, was published by Curtze, from fifteenth century MSS., in 1902[54]. Pages 130–159 of this edition contain "capitulum tertium in arearum divisionum explanatione" with Latin and German text, and among the many other propositions given by Savasorda is that of Proclus-Euclid (= Woepcke 28 = Leonardo 57). Compared with Leonardo's treatment of divisions Savasorda's seems rather trivial. But however great Leonardo's obligations to other writers, his originality and power sufficed to make a comprehensive and unified treatise.

Almost contemporary with Leonardo was Jordanus Nemorarius (d. 1237) who was the author of several works, all probably written before 1222. Among these is *Geometria vel De Triangulis*[55] in four books. The second book is principally devoted to problems on divisions: Propositions 1–7 to the division of lines and Propositions 8, 13, 17, 18, 19 to the division of rectilineal figures. The enunciations of Propositions 8, 13, 17, 19 correspond, respectively, to Euclid 3, 26, 19, 14 and to Leonardo 2, 4, 3, 36. But Jordanus's proofs are quite differently stated from those of Euclid or Leonardo. Both for themselves and for comparison with the Euclidean proofs which have come down to us, it will be interesting to reproduce propositions 13 and 17 of Jordanus.

"13. *Triangulo dato et puncto extra ipsum signato lineam per punctum transeuntem designare, que triangulum per equalia parciatur*" [pp. 15–16].

---

[53] That is, Abraham son of Chijja the prince. *Cf.* STEINSCHNEIDER, *Bibliotheca Mathematica*, 1896, (2), X, 34–38, and CANTOR, *Vorlesungen über Geschichte d. Math.* I_3, 797–800, 907.

[54] M. CURTZE, "Urkunden zur Geschichte der Mathematik im Mittelalter und der Renaissance..." Erster Teil (*Abhandlung zur Geschichte der Mathematischen Wissenschaften...* XII. Heft), Leipzig, 1902, pp. 3–183.

[55] Edited with Introduction by MAX CURTZE, *Mitteilungen des Copernicus-Vereins für Wissenschaften und Kunst zu Thorn.* VI. Heft, 1887. In his discussion of the second book, CANTOR (*Vorlesungen ü. Gesch. d. Math.* II_2, 75) is misleading and inaccurate. One phase of his inaccuracy has been referred to by ENESTRÖM (*Bibliotheca Mathematica*, Januar, 1912, (3), XII, 62).

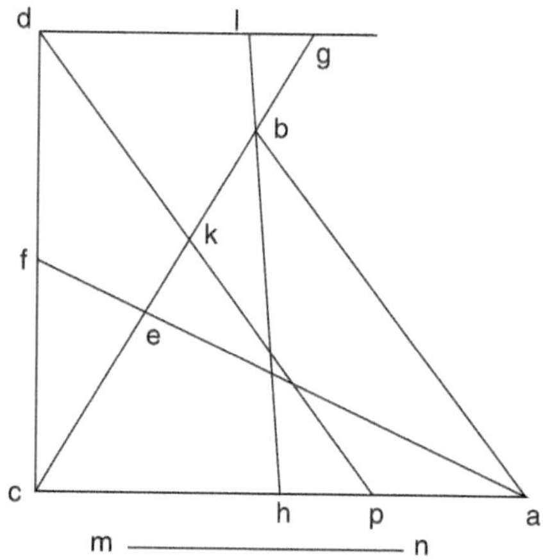

"Let abc be the triangle and d the point outside but contained within the lines aef, hbl, which are lines dividing the triangle equally and produced. For if d be taken in any such place, draw dg parallel to ca meeting cb produced in g. Join cd and find mn such that

$$\triangle cdg : \triangle aec \ (= \tfrac{1}{2} \triangle abc) = cg : mn.$$

Then divide cg in k such that

$$gk : kc = kc : mn.$$

Produce dk to meet ca in p. Then I say that dp divides the triangle abc into equal parts.

For, since the triangle ckp is similar to the triangle kdg, by 4 of sixth[56] and parallel lines and 15 of first and definitions of similar areas,

$$\triangle ckp : \triangle kdg = mn : kg$$

by corollary to 17 of sixth[57]. But

$$\triangle kdg : \triangle cdg = kg : cg.$$

Therefore, by equal proportions,

$$\triangle ckp : \triangle cdg = mn : cg.$$

$$\therefore \triangle ckp : \triangle cdg = \triangle aec : \triangle cdg.$$

And $$\triangle ckp = \triangle aec \ (= \tfrac{1}{2}\triangle abc)$$

[56] That is, Euclid's *Elements*, VI. 4.

[57] I do not know the MS. of Euclid here referred to; but manifestly it is the Porism of *Elements* VI. 19 which is quoted: "If three straight lines be proportional, then as the first is to the third, so is the figure described on the first to that which is similar and similarly described on the second."

by 9 of fifth, and this is the proposition.
And by the same process of deduction we may be led to an absurdity, namely, that all may equal a part if the point k be otherwise than between e and b or the point p be otherwise than between h and a; the part cut off must always be either all or part of the triangle aec."

"17. *Puncto infra propositum trigonum dato lineam per ipsum deducere, que triangulum secet per equalia*" [pp. 17–18].

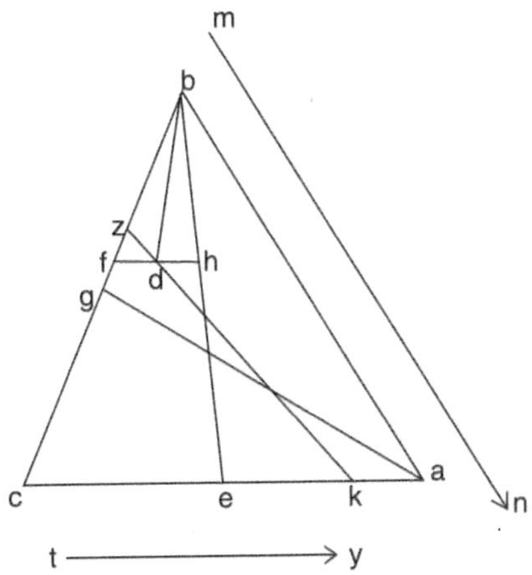

"Let abc be the triangle and d the point inside and contained within the part between ag and be which divide two sides and triangle into equal parts. Through d draw fdh parallel to ac and draw db. Then by 12 of this book[58] draw mn such that

$$bf : mn = \triangle bdf : \triangle bec \; (= \tfrac{1}{2}\triangle abc).$$

Also find ty such that

$$bf : ty = \triangle bfh : \triangle bec.$$

And since $\triangle bfh > \triangle bdf, \quad mn > ty$

by 8 and 10 of fifth.

Now $bf : bc = bc : ty$

58 That is, *De Triangulis*, Book 2, Prop. 12: "Data recta linea aliam rectam inuenire, ad quam se habeat prior sicut quilibet datus triangulus ad quemlibet datum triangulum" [p. 15].

by corollary to 17 of sixth[581], and $\triangle bfh < \triangle bec$ since fh, ce are parallel lines.

But
$$bc : ty > bc : mn$$

by second part of 8 of fifth.

$$\therefore bf : bc > bc : mn;$$
$$\therefore fc < \tfrac{1}{4} mn$$

by 6 of this book[59].

Add then to the line cf, from f, a line fz, by 5 of this book[60], such that

$$fz : zc = zc : mn;$$

and fz will be less than fb by the first part of the premise. [Supposition with regard to d?]

Join zd and produce it to meet ac in k; then I say that the line zdk divides the triangle abc into equal parts. For

$$\triangle bdf : \triangle zdf = bf : zf$$

by 1 of sixth.

But
$$\triangle zdf : \triangle zkc = zf : mn$$

by corollary to 17 of sixth[57] and similar triangles.

Therefore by 1 and by equal proportions

$$\triangle bdf : \triangle zkc = bf : mn.$$

But
$$\triangle bdf : \triangle bec = bf : mn.$$

Therefore by the second part of 9 of fifth

$$\triangle zkc = \triangle bec = \tfrac{1}{2} \triangle abc."$$ Q.E.F.

---

[581] Rather is it the converse of this corollary, which is quoted in note 57. It follows at once, however:

$$bf : ty = \triangle bfh : \triangle bec = bf^2 : bc^2, \quad \therefore bf \cdot ty = bc^2 \text{ or } bf : bc = bc : ty.$$

[59] "Cum sit linee breuiori adiecte major proporcio ad compositam, quam composite ad longiorem, breuiorem quarta longioris minorem esse necesse est [p. 13].

[60] "Duabus lineis propositis, quarum una sit minor quarta alterius uel equalis, minori talem lineam adiungere, ut, que adiecte ad compositam, eadem sit composite ad reliquam propositarum proporcio" [p. 12].

Proposition 18 of Jordanus is devoted to finding the centre of gravity of a triangle[601] and it is stated in the form of a problem on divisions. In Leonardo this problem is treated[109] by showing that the medians of a triangle are concurrent; but in Jordanus (as in Heron[83]) the question discussed is, "to find a point in a triangle such that when it is joined to the angular points, the triangle will be divided into three equal parts"(p. 18).

A much later work, *Summa de Arithmetica Geometria Proportioni et Proportionalita...* by Luca Paciuolo (b. about 1445) was published at Venice in 1494[61]. In the geometrical section (the second, and separately paged) of the work, pages 35 *verso*–43 *verso*, problems on divisions of figures are solved, and in this connection the author acknowledges great debt to Leonardo's work. Although the treatment is not as full as Leonardo's, yet practically the same figures are employed. The Proclus-Euclid propositions which have to do with the division of a circle are to be found here.

19. *"Muhammed Bagdedinus" and other Arabian writers on Divisions of Figures.*—We have not considered so far who "Muhammed Bagdedinus" was, other than to quote the statement of Dee[35] that he may have been "that *Albategnus* whom Copernicus often cites as a very considerable author, or that *Machomet* who is said to have been Alkindus's scholar." Albategnius or Muhammed b. Gâbir b. Sinân, Abû 'Abdallâh, el Battânî who received his name from Battân, in Syria, where he was born, lived in the latter part of the ninth and in the early part of the tenth century[62]. El-Kindî (d. about 873) the philosopher of the Arabians was in his prime about 850[63]. "Alkindus's scholar" would therefore possibly be a contemporary of Albategnius. It is probably because of these suggestions of Dee[64] that Chasles speaks[65] of "Mahomet Bagdadin, géomètre du xᵉ siècle."

It would be scarcely profitable to do more than give references to the recorded

601 Archimedes proved (*Works of Archimedes*, Heath ed., 1897, p. 201; *Opera omnia* iterum edidit J. L. Heiberg, II, 150–159, 1913) in Propositions 13–14, Book I of "On the Equilibrium of Planes" that *the centre of gravity of any triangle is at the intersection of the lines drawn from any two angles to the middle points of the opposite sides respectively.*

61 A new edition appeared at Toscolano in 1523, and in the section which we are discussing there does not appear to be any material change.

62 M. CANTOR, *Vorlesungen ü. Gesch. d. Math.* I₃, 736.

63 M. CANTOR, *Vorlesungen ü. Gesch. d. Math.* I₃, 718.

64 *Cf.* STEINSCHNEIDER[12].

65 CHASLES, *Aperçu historique...* 3ᵉ éd., Paris, 1889, p. 497.

opinions of other writers such as Smith[66], Kästner[67], Fabricius[68], Heilbronner[69], Montucla[70], Hankel[71], Grunert[72]— whose results Favaro summarizes[73].

The latest and most trustworthy research in this connection seems to be due to Suter who first surmised[74] that the author of the Dee book On Divisions was Muḥ. b. Muḥ. el-Baġdâdî who wrote at Cairo a table of sines for every minute. A little later[75], however, Suter discovered facts which led him to believe that the true author was Abû Muḥammed b. 'Abdelbâqî el-Baġdâdî (d. 1141 at the age of over 70 years) to whom an excellent commentary on Book X of the *Elements* has been ascribed. Of a MS. by this author Gherard of Cremona (1114–1187) may well have been a translator.

Euclid's book *On Divisions* was undoubtedly the ultimate basis of all Arabian works on the same subject. We have record of two or three other treatises.

1. Ṯâbit b. Qorra (826–901) translated parts of the works of Archimedes and Apollonius, revised Ishâq's translation of Euclid's *Elements* and *Data* and also revised the work *On Divisions of Figures* translated by an anonymous writer[76].

2. Abû Muḥ. el-Hasan b. 'Obeidallâh b. Soleimân b. Wahb (d. 901) was a distinguished geometer who wrote "A Commentary on the difficult parts of the work of Euclid" and "The Book on Proportion." Suter thinks[77] that another reading is possible in connection with the second title, and that it may refer to Euclid's work *On Divisions.*

3. Abû'l Wefâ el-Bûzǧânî (940–997) one of the greatest of Arabian mathematicians and astronomers spent his later life in Bagdad, and is the author of a course of Lectures on geometrical constructions. Chapters VII–IX of the Persian form of this treatise which has come down to us in roundabout fashion were entitled: "On the division of triangles," "On the division of quadrilaterals," "On

---

66 T. Smith, *Vitae quorumdam... virorum*, 1707, p. 56. *Cf.* notes 14, 15.

67 A. G. Kästner, *Geschichte der Mathematik...*, Band I, Göttingen, 1796, p. 273. See also his preface to N. Morville, *Lehre von der geometrischen und ökonomischen Vertheilung der Felder, nach der dänischen Schrift bearbeitet von J. W. Christiani, begleitet mit einer Vorrede... von A. G. Kästner*, Göttingen, 1793.

68 J. A. Fabricius, *Bibliotheca Graeca...* Editio nova. Volumen quartum, Hamburgi, MDCCLXXXV, p. 81.

69 J. C. Heilbronner, *Historia Matheseos universae...* Lipsiae, MDCCXLII, p. 438, 163–4.

70 J. F. Montucla, *Histoire des mathématiques...* éd. nouv. Tome I, An VII, p. 216.

71 H. Hankel, *Zur Geschichte der Math. in Alterthum u. Mittelalter*, Leipzig, 1874, p. 234.

72 J. A. Grunert, *Math. Wörterbuch ... von G.S. Klügel, fortgesetzt von C. B. Mollweide und beendigt von J.A. Grunert...* Erste Abteilung, die reine Math., fünfter Theil, erster Band, Leipzig, 1831, p. 76.

73 Favaro, pp. 141–144.

74 H. Suter, "Die Mathematiker und Astronomen der Araber und ihre Werke" (*Abh. z. Gesch. d. Math. Wiss.* X. Heft, Leipzig, 1900), p. 202, No. 517.

75 H. Suter, *idem*, "Nachträge und Berichtigung" (*Abh. z. Gesch. d. Math. Wiss.* XIV. Heft, 1902), p. 181; also *Bibliotheca Mathematica*, IV$_3$, 1903, pp. 22–27.

76 H. Suter, "Die Mathematiker...," pp. 34–38.

77 H. Suter, "Die Mathematiker...," pp. 48 and 211, note 23.

the division of circles" respectively. Chapter VII and the beginning of Chapter VIII are, however, missing from the Bibliothèque nationale Persian MS. which has been described by Woepcke[78]. This MS., which gives constructions without demonstrations, was made from an Arabian text, by one Abû Ishâq b. 'Abdallâh with the assistance of four pupils and the aid of another translation. The Arabian text was an abridgment of Abû'l Wefâ's lectures prepared by a gifted disciple.

The three propositions of Chapter IX[79] are practically identical with Euclid (Woepcke) 28, 29. In Chapter VIII[80] there are 24 propositions. About a score are given, in substance, by both Leonardo and Euclid.

In conclusion, it may be remarked that in Chapter XII of Abû'l Wefâ's work are 9 propositions, with various solutions, for dividing the surface of a sphere into equiangular and equilateral triangles, quadrilaterals, pentagons and hexagons.

20. *Practical applications of the problems On Divisions of Figures; the μετρικά of Heron of Alexandria.*—The popularity of the problems of Euclid's book *On Divisions* among Arabians, as well as later in Europe, was no doubt largely due to the possible practical application of the problems in the division of parcels of land of various shapes, the areas of which, according to the Rhind papyrus, were already discussed in empirical fashion about 1800 B. C. In the first century before Christ[81] we find that Heron of Alexandria dealt with the division of surfaces and solids in the third book of his *Surveying* (*μετρικά*)[82]. Although the enunciations of the propositions in this book are, as a whole, similar[83] to those in Euclid's book *On Divisions*, Heron's discussion consists

[78] F. WOEPCKE, "Recherches sur l'histoire des Sciences mathématiques chez les orientaux, d'après des traités inedits Arabes et Persans. Deuxième article. Analyse et extrait d'un recueil de constructions géométriques par Aboûl Wafâ," *Journal asiatique*, Fevrier–Avril, 1855, (5), V, 218–256, 309–359; reprint, Paris, 1855, pp. 89.

[79] F. WOEPCKE, *idem*, pp. 340–341; reprint, pp. 70–71.

[80] F. WOEPCKE, *idem*, pp. 338–340; reprint, pp. 68–70.

[81] This date is uncertain, but recent research appears to place it not earlier than 50 B. C. nor later than 150 A. D. *Cf.* HEATH, *Thirteen Books of Euclid's Elements*, I, 20–21; or perhaps better still, Article "Heron 5" by K. Tittel in Pauly-Wissowa's *Real-Encyclopädie der class. Altertumswissenschaften*, VIII, Stuttgart, 1913, especially columns 996–1000.

[82] *Heronis Alexandrini opera quae supersunt omnia*, Vol. III, *Rationes Dimetiendi et commentatio Dioptrica recensuit Hermannus Schoene*, Lipsiae, MCMIII. Third book, pp. 140–185. *Cf.* CANTOR, *Vorlesungen*..., I_3, 380–382.

[83] Only two are exactly the same: II–III (= Euclid 30), VII (= Euclid 32), the problem considered in X is practically Euclid 27 (Art. 48), while XVIII is closely related to Euclid 29 (Art. 50). In XIX Heron finds in a triangle a point such that when it is joined to the angular points, the triangle will be divided into three equal parts. The divisions of solids of which Heron treats are of a sphere (XXIII) and the division in a given ratio, by a plane parallel to the base, of a Pyramid (XX) and of a Cone (XXI). For proof of Proposition XXIII: *To cut a sphere by a plane so that the volumes of the segments are to one another in a given ratio*, Heron refers to Proposition 4, Book II of "On the Sphere and Cylinder" of Archimedes; the third proposition in the same book

almost entirely of "analyses" and approximations. For example, II: "To divide a triangle in a given ratio by a line drawn parallel to the base"—while Euclid gives the general construction, Heron considers that the sides of the given triangle have certain known numerical lengths and thence finds the approximate distance of the angular points of the triangle to the points in the sides where the required line parallel to the base intersects them, because, as he expressly states, in a field with uneven surface it is difficult to draw a line parallel to another. Most of the problems are discussed with a variety of numbers although theoretical analysis sometimes enters. Take as an example Proposition X[84]: "*To divide a triangle in a given ratio by a line drawn from a point in a side produced.*"

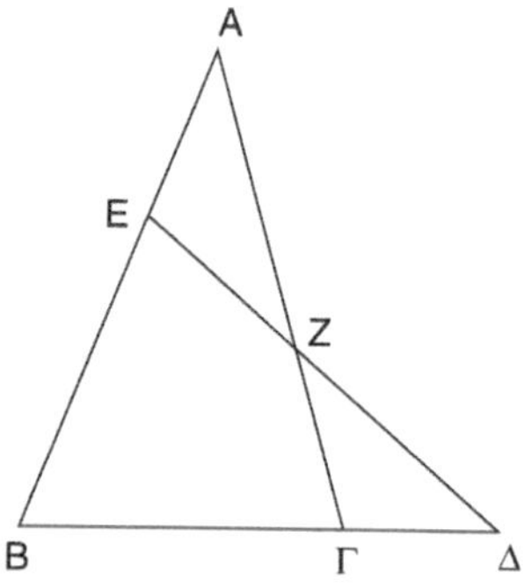

"Suppose the construction made. Then the ratio of triangle AEZ to quadrilateral $ZEB\Gamma$ is known; also the ratio of the triangle $AB\Gamma$ to the triangle AZE. But the triangle $AB\Gamma$ is known, therefore so is the triangle AZE. Now Δ is given. Through a known point Δ there is therefore drawn a line which, with two lines AB and $A\Gamma$ intersecting in A, encloses a known area.

Therefore the points E and Z are given. This is shown in the second book of *On Cutting off a Space*. Hence the required proof.

If the point Δ be not on $B\Gamma$ but anywhere this will make no difference."

21. *Connection between Euclid's book On Divisions, Apollonius's treatise On Cutting off a Space and a Pappus-lemma to Euclid's book of Porisms.*—Although the name of the author of the above-mentioned work is not given by

---

of the Archimedean work is (Heron XVII): *To cut a given sphere by a plane so that the surfaces of the segments may have to one another a given ratio.* (*Works of Archimedes*, Heath ed., 1897, pp. 61–65; *Opera omnia* iterum edidit J. L. Heiberg, I, 184–195, 1910.)

Propositions II and VII are also given in Heron's περὶ διόπτρας (Schoene's edition, pp. 278–281). *Cf.* "Extraits des Manuscrits relatifs à la géométrie grecs" par A. J. C. Vincent, *Notices et extraits des Manuscrits de la bibliothèque impériale*, Paris, 1858, XIX, pp. 157, 283, 285.

[84] HERON, *idem*, p. 160f.

Heron, the reference is clearly to Apollonius's lost work. According to Pappus it consisted of two books which contained 124 propositions treating of the various cases of the following problem: *Given two coplanar straight lines* A_1P_1, B_2P_2, *on which* A_1 *and* B_2 *are fixed points; it is required to draw through a fixed point* Δ *of the plane, a transversal* ΔZE *forming on* A_1P_1, B_2P_2 *the two segments* A_1Z, B_2E *such that* $A_1Z \cdot B_2E$ *is equal to a given rectangle.*

Given a construction for the particular case when A_1P_1, B_2P_2 meet in A, and when A_1 and B_2 coincide with A—Heron's reasoning becomes clear. The solution of this particular case is practically equivalent to the solution of Euclid's Proposition 19 or 20 or 26 or 27. References to restorations of Apollonius's work are given in note 111.

To complete the list of references to writers before 1500, who have treated

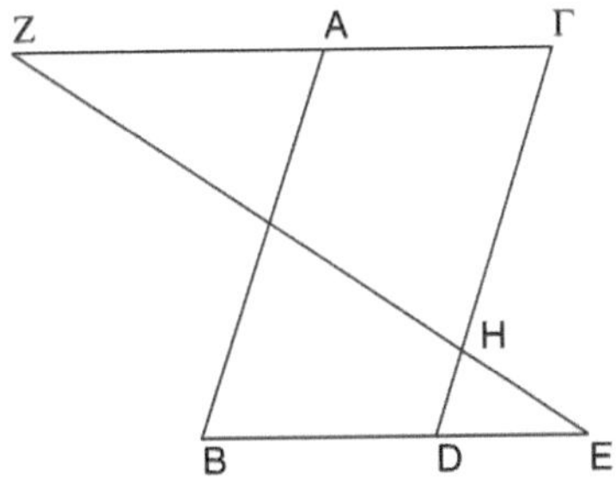

of Euclid's problems here under discussion, I should not fail to mention the last of the 38 lemmas which Pappus gives as useful in connection with the 171 theorems of Euclid's lost book of *Porisms: Through a given point* E *in* BD *produced to draw a line cutting the parallelogram* AD *such that the triangle* $Z\Gamma H$ *is equal to the parallelogram* AD.

After "Analysis" Pappus has the following

"Synthesis. Given the parallelogram AD and the point E. Through E draw the line EZ such that the rectangle $\Gamma Z \cdot \Gamma H$ equals twice the rectangle $A\Gamma \cdot \Gamma D$. Then according to the above analysis [which contains a reference to an earlier lemma discussed a little later[88] in this book] the triangle $Z\Gamma H$ equals the parallelogram AD. Hence EZ satisfies the problem and is the only line to do so[85]."

The tacit assumption here made, that the equivalent of a proposition of Euclid's book *On Divisions (of Figures)* was well known, is noteworthy.

[85] Pappus ed. by Hultsch, Vol. 2, Berlin, 1877, pp. 917–919. In Chasles's restoration of Euclid's *Porisms*, this lemma is used in connection with "Porism CLXXX: Given two lines SA, SA', a point P and a space ν: points I and J' can be found in a line with P and such that if one take on SA, SA' two points m, m', bound by the equation $Im \cdot J'm' = \nu$, the line mm' will pass through a given point." *Les trois livres de Porismes d'Euclide*, Paris, 1860, p. 284. See also the restoration by R. Simson, pp. 527–530 of "De porismatibus tractatus," *Opera quaedam reliqua...* Glasguae, M.DCC.LXXVI.

III.

"*The Treatise of Euclid on the Division (of plane Figures).*"

Proposition 1.

22. "*To divide*[86] *a given triangle into equal parts by a line parallel to its base.*" [Leonardo 5, p. 119, ll. 7–9.]

Let abg be the given triangle which it is required to bisect by a line parallel to bg. Produce ba to d till $ba = 2ad$. Then in ba find a point e such that

$$ba : ae = ae : ad.$$

Through e draw ez parallel to bg; then the triangle abg is divided by the line ez into two equal parts, of which one is the triangle aez, and the other the quadrilateral $ebgz$.

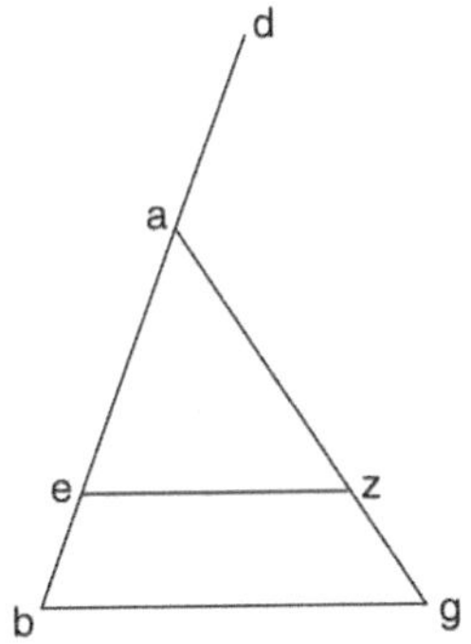

Leonardo then gives three proofs, but as the first and second are practically equivalent, I shall only indicate the second and third.

I. When three lines are proportional, as the first is to the third so is a figure on the first to the similar and similarly situated figure described on the second [VI. 19, "Porism"][87].

$\therefore$ $ba : ad$ = figure on ba : similar and similarly situated figure on ae.

---

[86] Literally, the original runs, according to Woepcke, "We propose to ourselves to demonstrate how to divide, etc." I have added all footnotes except those attributed to Woepcke.

[87] Throughout the restoration I have added occasional references of this kind to Heath's edition of Euclid's *Elements*; VI. 19 refers to Proposition 19 of Book VI. *Cf.* note 57.

Hence $$ba : ad = \triangle abg : \triangle aez$$
$$= 2 : 1.$$

$$\therefore \triangle abg = 2 \triangle aez.$$

II. $$ba : ae = ae : ad.$$
$$\therefore ba \cdot ad = ae^2,$$

and since ad is one-half of ba,

$$ba^2 = 2\,ae^2.$$

And since bg is parallel to ez,

$$ba : ae = ga : az.$$
$$\therefore ba^2 : ae^2 = ga^2 : az^2. \qquad \text{[VI. 22]}$$

But $$ba^2 = 2\,ae^2.$$
$$\therefore ga^2 = 2\,az^2.$$

Then $$ba \cdot ag = 2\,ae \cdot az, \qquad \text{[VI. 22]}$$
$$\therefore \triangle abg = 2 \triangle aez^{88}.$$

Then follows a numerical example.

[88] The theorem here assumed is enunciated by Leonardo (p. 111, ll. 24–27) as follows:

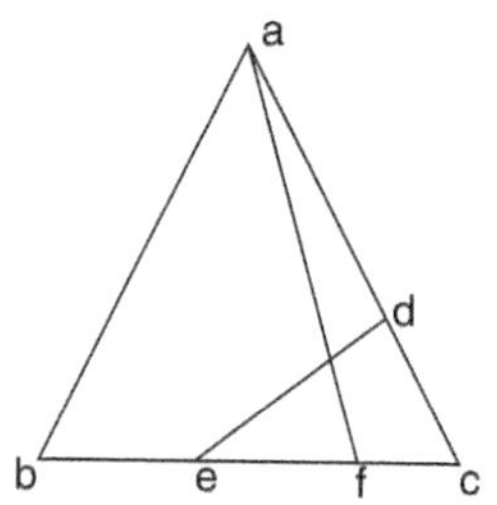

Et si á trigono recta protracta fuerit secans duo latera trigonj, que cum ipsis duobus lateribus faciant trigonum habentem angulum unum comunem cum ipso trigono, erit proportio unius trigoni ad alium, sicut facta ex lateribus continentibus ipsum angulum. This is followed by the sentence "Ad cuius rei euidentiam." Then come the construction and proof:

Let abc be the given triangle and de the line across it, meeting the sides ca and cb in the points d, e, respectively. I say that

$$\triangle abc : \triangle dec = ac \cdot cb : dc \cdot ce.$$

Proof: To ac apply the triangle $afc = \triangle dec$. [I. 44]
Since the triangles abc, afc are of the same altitude,

$$bc : fc = \triangle abc : \triangle afc. \qquad [\text{VI. } 1]$$

But

$$bc : fc = ac \cdot bc : ac \cdot fc, \qquad [\text{V. } 15]$$

$$\therefore \triangle abc : \triangle afc = ac \cdot bc : ac \cdot fc,$$

and since $\triangle dec = \triangle acf$,

$$\triangle acb : \triangle dce = ac \cdot bc : ac \cdot cf.$$

Again, since the triangles acf, dce are equal and have a common angle, as in the fifteenth theorem of the sixth book of Euclid, the sides are mutually proportional.

$$\therefore ac : dc = ce : cf, \qquad \therefore ac \cdot cf = dc \cdot ce,$$

$$\therefore \triangle acb : \triangle dce = ac \cdot cb : dc \cdot ce.$$

"quod oportebat ostendere."

It is to be observed that the Latin letters are used with the above figure. This suggests the possibility of the proof being due to Leonardo.

The theorem is assumed in Euclid's proof of proposition 19 (Art. 40) and it occurs, directly or indirectly, in more than one of his works. A proof, depending on the proposition that the area of a triangle is equal to one-half the product of its base and altitude, is given by Pappus (pp. 894–897) in connection with one of his lemmas for Euclid's book of *Porisms*: *Triangles which have one angle of the one equal or supplementary to one angle of the other are in the ratio compounded of the ratios of the sides about the equal or supplementary angles.* (*Cf.* R. SIMSON, "De Porismatibus Tractatus" in *Opera quaedam reliqua...* 1776, p. 515 ff.—P. BRETON (de Champ), "Recherches nouvelles sur les porismes d'Euclide," *Journal de mathématiques pures et appliquées*, XX, 1855, p. 233 ff. Reprint, p. 25 ff.—M. CHASLES, *Les trois livres de Porismes d'Euclide...* Paris, 1860, pp. 247, 295, 307.)

The first part of this lemma is practically equivalent to either (1) [VI. 23]: *Equiangular parallelograms have to one another the ratio compounded of the ratio of their sides*; or (2) the first part of Prop. 70 of the *Data* (*Euclidis Data...* edidit H. Menge, Lipsiae, 1896, p. 130f.): *If in two equiangular parallelograms the sides containing the equal angles have a given ratio to one another* [i. e. one side in one to one side in the other], *the parallelograms themselves will also have a given ratio to one another. Cf.* HEATH, *Thirteen Books of Euclid's Elements*, II, 250.

The proposition is stated in another way by Pappus[85] (p. 928) who proves that *a parallelogram is to an equiangular parallelogram as the rectangle contained by the adjacent sides of the first is to the rectangle contained by the adjacent sides of the second.*

The above theorem of Leonardo is precisely the first of those theorems which Commandinus adds to VI. 17 of *his* edition of Euclid's *Elements* and concerning which he writes "à nobis elaborata" ("fatti da noi"): *Euclidis Elementorum Libri XV... A Federico Commandino...* Pisauri, MDLXXII, p. 81 *recto (Degli Elementi d' Euclide libri*

PROPOSITION 2.

23. *"To divide a given triangle into three equal parts by two lines parallel to its base."* [Leonardo 14, p. 122, l. 8.]

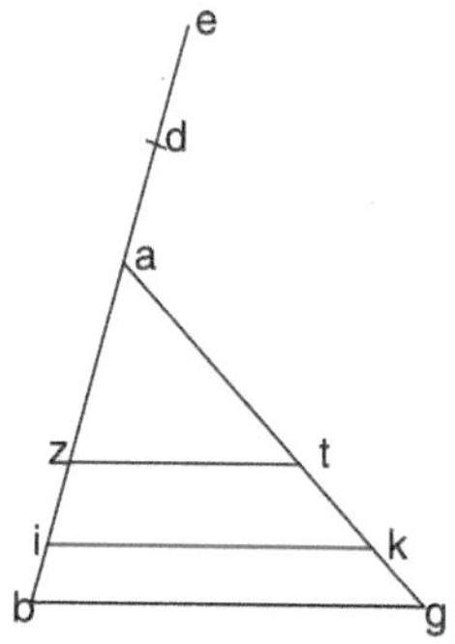

Let abg be the given triangle with base bg. Produce ba to d till $ba = 3ad$, and produce ad to e till $ad = de$; then $ae = \frac{2}{3}ba$. Find az, a mean proportional between ba and ad, and ia a mean proportional between ba and ae. Then through z and i draw zt, ik parallel to bg and I say that the triangle abg is divided into three equal parts of which one is the triangle azt, another the quadrilateral $zikt$, the third the quadrilateral $ibgk$.

Proof: Since

$$ba : az = az : ad,$$

$$ba : ad = \triangle abg : \triangle azt, \qquad \text{[VI. 19, Porism]}$$

for these triangles are similar.

Now $\qquad ba = 3ad; \quad \therefore \triangle abg = 3\triangle azt.$

$$\therefore \triangle azt = \tfrac{1}{3}\triangle abg.$$

Again, $\qquad ba : ia = ia : ae;$

$\therefore ba : ae = \triangle$ on ea: similar and similarly situated $\triangle$ on ai.

But triangles aik, abg are similar and similarly described on ai and ab; and

$$ea : ab = 2 : 3.$$

$$\therefore \triangle aik = \tfrac{2}{3}\triangle abg.$$

---

quindici con gli scholii antichi tradotti prima in lingua latina da M. Federico Commandino *da Urbino, et con commentarii illustrati, et hora d' ordine dell' istesso transportati nella nostra vulgare, et da lui riveduti.* In Urbino, M.D.LXXV, p. 88 *recto*).

And since $\triangle azt = \frac{1}{3}\triangle abg$, there remains the quadrilateral $zikt = \frac{1}{3}\triangle abg$. We see that the quadrilateral $ibgk$ will be the other third part; hence the triangle abg has been divided into three equal parts; "quod oportebat facere."

Leonardo continues: "Et sic per demonstratos modos omnia genera trigonorum possunt diuidi in quatuor partes uel plures." Cf. note 45.

PROPOSITION 3.

24. *"To divide a given triangle into two equal parts by a line drawn from a given point situated on one of the sides of the triangle."* [Leonardo 1, 2, p. 110, l. 31; p. 111, ll. 41–43.]

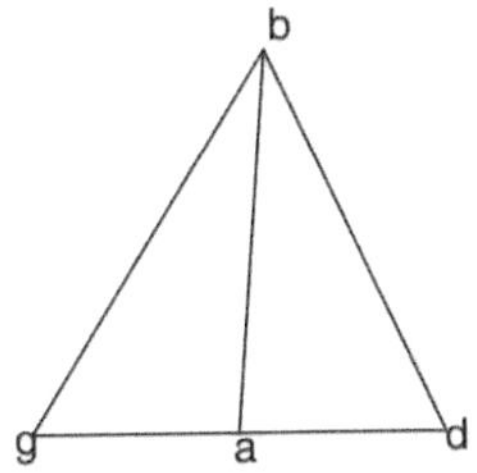

Given the triangle bgd; if a be the middle point of gd the line ba will divide the triangle as required; either because the triangles are on equal bases and of the same altitude [I. 38; Leonardo 1], or because

$$\triangle bgd : \triangle bad = bd \cdot dg : bd \cdot da^{88}.$$

Whence
$$\triangle bgd = 2\triangle bad.$$

But if the given point be not the middle point of any side, let abg be the triangle and d the given point nearer to b than to g. Bisect bg at e and draw ad, ae. Through e draw ez parallel to da; join dz. Then the triangle abg is bisected by dz.

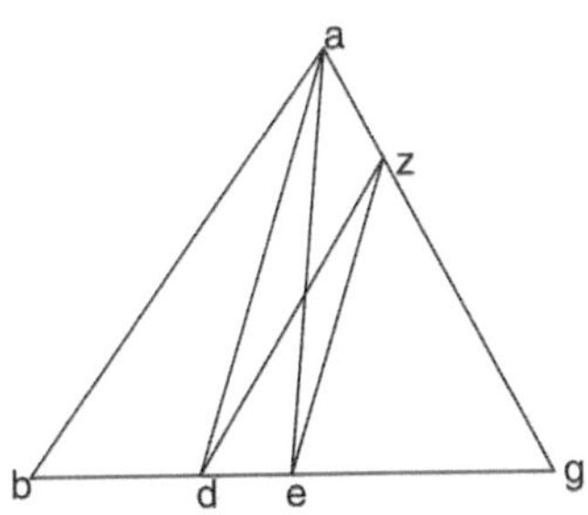

Proof: Since

$$ad \parallel ez, \quad \triangle adz = \triangle ade.$$

To each add $\triangle abd$. Then

$$\text{quadl.}abdz = \triangle abd + \triangle ade,$$
$$= \triangle abe.$$

But

$$\triangle abe = \tfrac{1}{2}\triangle abg;$$
$$\therefore \text{quadl.}abdz = \tfrac{1}{2}\triangle abg;$$

and the triangle zdg is the other half of the triangle abg. Therefore the triangle abg is divided into two equal parts by the line dz drawn from the point d;

"ut oportebat facere."

Then follows a numerical example.

Proposition 4.

25. *"To divide a given trapezium*[89] *into two equal parts by a line parallel to its base."* [Leonardo 23, p. 125, ll. 37–38.]

Let $abgd$ be the given trapezium with parallel sides ad, bg, ad being the lesser. It is required to bisect the trapezium by a line parallel to the base bg. Let gd, ba, produced, meet in a point e. Determine z such that

$$ze^2 = \tfrac{1}{2}(eb^2 + ea^2).^{90}$$

Through z draw zi parallel to gb. I say that the trapezium $abgd$ is divided into two equal parts by the line zi parallel to the base bg.

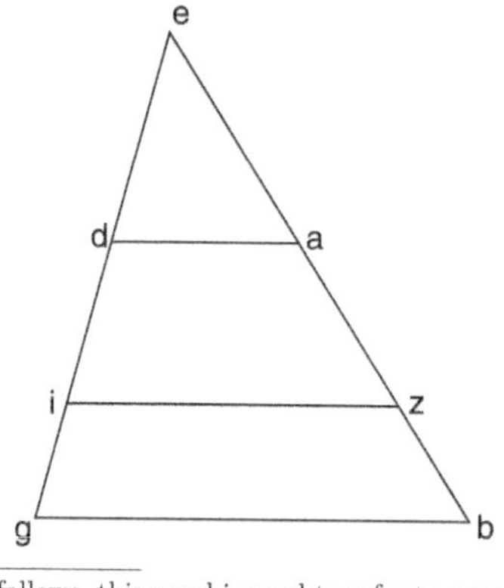

[89] Here, and in what follows, this word is used to refer to a quadrilateral two of whose sides are parallel.

[90] The point z is easily found by constructions which twice make use of I. 47.

Proof: For since

$$2ze^2 = eb^2 + ea^2,$$

and all the triangles are similar,

$$2\triangle ezi = \triangle ebg + \triangle ead. \qquad \text{[VI. 19]}$$

From the triangle ebg take away the triangle ezi. Then

$$\triangle ezi = \text{quadl.} zbgi + \triangle eda.$$

And taking away from the equals the triangle eda, we get

$$\text{quadl.} ai = \text{quadl.} zg.$$

Therefore the trapezium $abgd$ is divided into two equal parts by the line zi parallel to its base. Q. O. F.

A numerical example then follows.

PROPOSITION 5.

26. *"And we divide the given trapezium into three equal parts as we divide the triangle, by a construction analogous to the preceding construction*[91]*."* [Leonardo 33, p. 134, ll. 14–15.]

Let $abgd$ be the trapezium with parallel sides ad, bg and other sides ba, gd produced to meet in e. Let zti be a line such that

$$zi : it = eb^2 : ea^2.$$[92]

[91] It is to be noticed that Leonardo's discussion of this proposition is hardly "analogous to the preceding construction" which is certainly simpler than if it had been similar to that of Prop. 5. A construction for Prop. 4 along the same lines, which may well have been Euclid's method, would obviously be as follows:

Let zti be a line such that

$$zi : it = eb^2 : ea^2.$$

Divide tz into two equal parts, tk, kz. Find m such that

$$em^2 : eb^2 = ki : zi.$$

Then m leads to the same solution as before. [For, in brief,

$$em^2 = eb^2\left(\frac{ki}{zi}\right) = eb^2\left(\frac{\frac{zt}{2}+ti}{zi}\right) = \frac{eb^2}{2}\left(\frac{zi+it}{zi}\right) = \frac{eb^2}{2}\left(\frac{ea^2+eb^2}{eb^2}\right)$$
$$= \tfrac{1}{2}\left(ea^2+eb^2\right).]$$

[92] From VI. 19, Porism, it is clear that the construction here is to find a line x which is a third proportional to eb and ea. Then $zi : it = eb : x$.

Divide tz into three equal parts tk, kl, lz.
Find m and n in be such that

$$em^2 : eb^2 = ik : zi,$$

and

$$en^2 : eb^2 = il : zi.$$

Through m and n draw mo, np parallel to the base bg. Then I say that the quadrilateral ag is divided into three equal parts: ao, mp, ng.

Proof: For $eb^2 : ea^2 = \triangle ebg : \triangle ead.$ [VI. 19]

$\therefore zi : it = \triangle ebg : \triangle ead.$ .... [1]

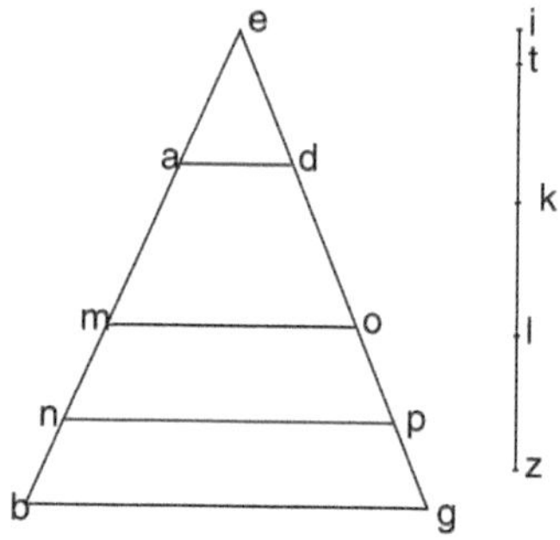

But $zi : ik = eb^2 : em^2,$

$\therefore zi : ik = \triangle ebg : \triangle emo.$ ...[2]

So also $zi : il = \triangle ebg : \triangle enp.$ ...[3]

Whence $it : tk = \triangle ead : \text{quadl.}ao,$[93]

and therefore $tk : kl = \text{quadl.}ao : \text{quadl.}mp.$[94]

But $tk = kl. \quad \therefore \text{quadl.}ao = \text{quadl.}mp.$

So also $kl : lz = \text{quadl.}mp : \text{quadl.}ng;$

and $kl = lz. \quad \therefore \text{quadl.}mp = \text{quadl.}ng.$

---

93 This may be obtained by combining [1] and [2], and applying V. 11, 16, 17.

94 Relations [1], [2] and [3] may be employed, as in the preceding, to give,

$$it : kl = \triangle ead : \text{quadl.}mp;$$

combining this with $it : tk = \triangle ead : \text{quadl.}ao$, we get the required result,

$$tk : kl = \text{quadl.}ao : \text{quadl.}mp.$$

Therefore the quadrilateral is divided into equal quadrilaterals *ao*, *mp*, *ng*; "ut prediximus."

Then follows a numerical example.

PROPOSITION 6.

27. *"To divide a parallelogram into two equal parts by a straight line drawn from a given point situated on one of the sides of the parallelogram."* [Leonardo 16, p. 123, ll. 30–31.]

Let *abcd* be the parallelogram and *i* any point in the side *ad*. Bisect *ad*

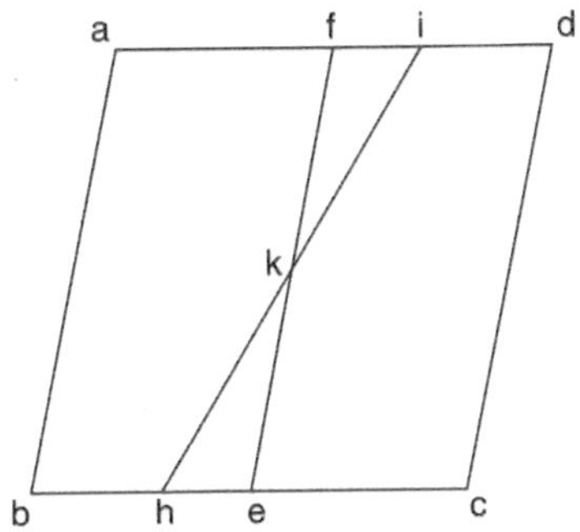

in *f* and *bc* in *e*. Join *fe*. Then the parallelogram *ac* is divided into equal parallelograms *ae*, *fc* on equal bases.

Cut off $eh = fi$. Join *hi*. Then this is the line required.

Leonardo gives two proofs:

I. Let *hi* meet *fe* in *k*. Then [$\triangle$s *fki*, *hke* are equal; add to each the pentagon *kfabh*, etc.]

II. Since *ae*, *fc* are $\square$s, $af = be$ and $fd = ec$. But

$$fd = \tfrac{1}{2}ad.$$
$$\therefore fd = af = ec.$$

And since $fi = he, \quad ai = ch.$

So also $di = bh$, and *hi* is common.

$$\therefore \text{quadl.}\, iabh = \text{quadl.}\, ihcd.$$[95]

[95] The first rather than the second proof is Euclidean. There is no proposition of the *Elements* with regard to the equality of quadrilaterals whose sides and angles, taken in the same order, are equal. Of course the result is readily deduced from I. 4, if we make certain suppositions with regard to order. Cf. the proof of Prop. 10.

Similarly if the given point were between a and f, [etc.; or on any other side]. And thus a parallelogram can be divided into two equal parts by a straight line drawn from a given point situated on any one of its sides.

PROPOSITION 7.

28. "*To cut off a certain fraction from a given parallelogram by a straight line drawn from a given point situated on one of the sides of the parallelogram.*" [Leonardo 20 (the case where the fraction is one-third), p. 124, ll. 24–26.]

Let $abcd$ be the given parallelogram. Suppose it be required to cut off a third of this parallelogram, by a straight line drawn from i, in the side ad.

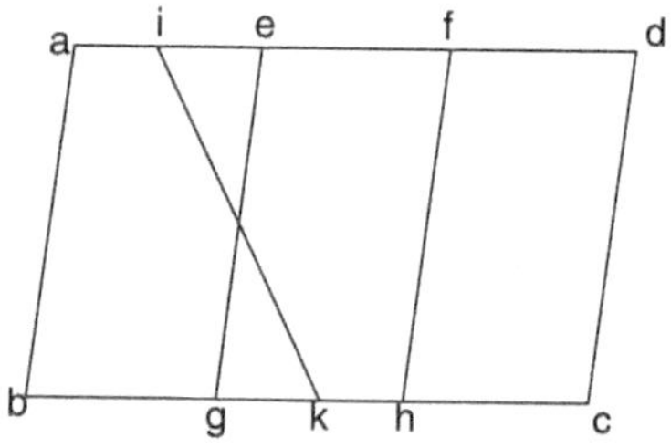

(The figure here is a combination of two in the original.)

Trisect ad in e and f and through e, f draw eg, fh parallel to dc; [then these lines trisect the ▭. If the point i be in the line ad, at either e or f, then the problem is solved. But if it be between a and e, draw ik to bisect the ▭ ah (Prop. 6), etc. Similarly if i were between e and f, or between f and d].

After finishing these cases Leonardo concludes:

"eodem modo potest omne paralilogramum diuidi in quatuor uel plures partes equales[45]."

The construction in this proposition is limited to the case where "a certain fraction" is the reciprocal of an integer. But more generally, if the fraction were $m : n$ (the ratio of the lengths of two given lines), we could proceed in a very similar way: Divide ad in e, internally, so that $ae : ed = m : n - m \quad (n > m)$. In ad cut off $ef = ae$ and through f draw fh parallel to ab. Then, as before, the problem is reduced to Proposition 6.

If the point e should fall at i or in the interval ai the part cut off from the parallelogram by the required line would be in the form of a triangle which might be determined by I. 44.

PROPOSITION 8.

29. *"To divide a given trapezium into two equal parts by a straight line drawn from a given point situated on the longer of the sides of the trapezium."* [Part of Leonardo 27[96], p. 127, ll. 2–3.]

This enunciation means, apparently, "from a given point situated on the longer of the [parallel] sides." At any rate Leonardo gives constructions for the cases when the given point is on any side. These I shall take up successively. The figure is made from more than one of Leonardo's, and there is a slight change in the lettering.

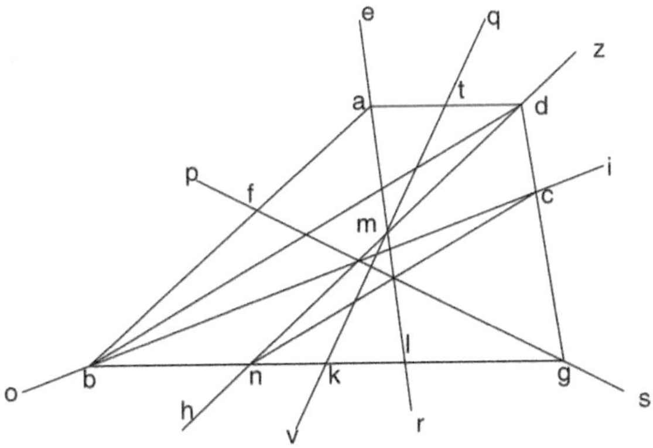

Let ad be the shorter of the parallel sides ad, bg, which are bisected in t and k respectively. Join tk. Then if bt, gt be joined, [it is clear, from triangles on equal bases and between the same parallels, that tk bisects the trapezium]. [This is Leonardo 24, p. 126, l. 31.]

Next consider the given point as any point on the shorter side [Leonardo 25, p. 127, ll. 2–3].

First let the point be at the angle a. Cut off kl in kg, equal to at. Join al, meeting tk in m; then the quadrilateral is divided as required by al. For [the triangles atm, mkl are equal in all respects, etc.].

Similarly if d were the given point; in kb cut off kn equal to td, and dn divides the quadrilateral into two equal parts which is proved as in the preceding case.

[Were the given point anywhere between a and t the other end of the bisecting line would be between k and l. Similarly if the given point were between t and d, the corresponding point would be between k and n.]

96 Leonardo 27: "Quomodo quadrilatera duorum laterum equidistantium diuidantur á puncto dato super quodlibet latus ipsius" [p. 129, ll. 2–3]. *Cf.* note 46.

Although not observed by Favaro, Leonardo now considers:

If the given point be in the side bg; either l, or n, or a point between l and n, then the above construction is at once applicable.

Suppose, however, that the given point were at b or in the segment bn, at g or in the segment lg. First consider the given point at b. Join bd and through n draw nc parallel to bd to meet gd in c. Join bc. Then bc bisects the trapezium. For [$abnd$ is half of the trapezium ag, and the triangle bnd equals the triangle bdc etc.].

Similarly from a given point between b and n, a line could be drawn meeting gd between c and d, and dividing the quadrilateral into two equal parts.

So also from g a line gf could be drawn [etc.]; and similarly for a given point between g and l.

Leonardo then concludes (p. 127, ll. 37–40):

"Jam ostensum est quomodo in duo equa quadrilatera duorum equidistantium laterum diuidi debeant á linea protracta ab omni dato puncto super lineas equidistantes ipsius; nunc uero ostendamus quomodo diuidantur á linea egrediente á dato puncto super reliqua latera."

This is overlooked by Favaro, though implied in his 27 [Leonardo, p. 129, l. 4]. I may add Leonardo's discussion of the above proposition although it does not seem to be called for by Euclid.

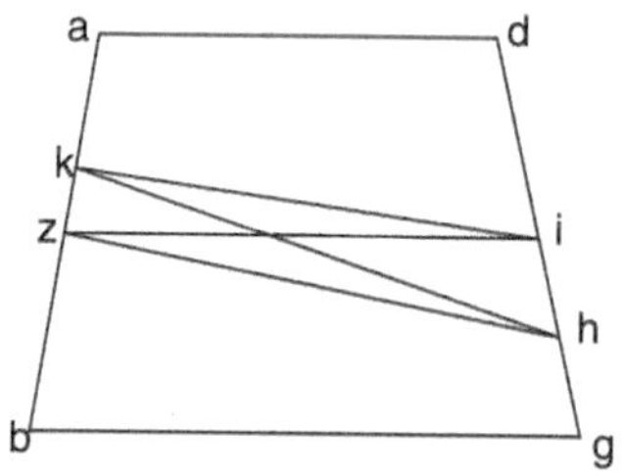

Let the point be in the side gd. For g or c or d or any point between c and d the above constructions clearly suffice. Let us, then, now consider the given point h as between c and g. Draw the line iz parallel to gb to bisect the trapezium (Prop. 4). Suppose h were between g and i. Join zh. Through i draw ik parallel to hz, and meeting ab in k.

(The lettering of the original figure is somewhat changed.)

Join hk, then [this is the line required; since

$$\triangle izh = \triangle kzh, \text{ etc.}].$$

[Similarly if h were between i and d.]

[So also for points on the line ab.]

Proposition 9.

30. *"To cut off a certain fraction from a given trapezium by a straight line drawn from a given point situated on the longer side of the trapezium."* [Leonardo 30, 31[97], p. 133, ll. 17–19, 31.]

I shall interpret "longer side" as in Proposition 8, and lead up to the consideration of any given point on bg after discussing the cases of points on the shorter side ad.

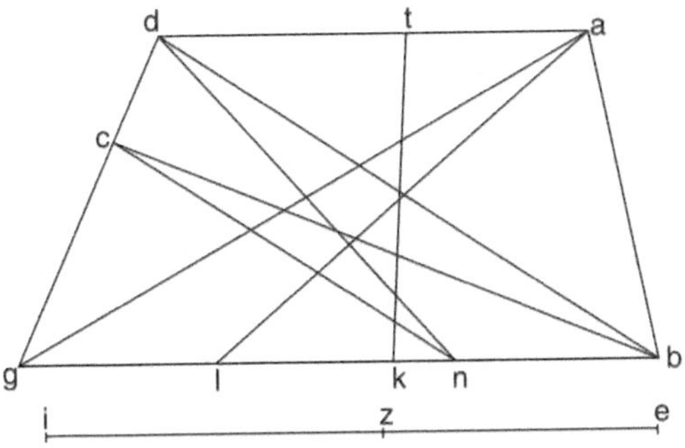

(This figure is made from three of Leonardo's.)

Suppose it be required to divide the trapezium in the ratio $ez : zi$[98].

Divide ad, bg in the points t, k, respectively, such that

$$at : td = ez : zi = bk : kg.$$

Join tk. Then by joining bt and gt [it is easily seen by VI. 1 and V. 12, that the trapezium ag is divided by tk in the ratio $ez : zi$].

If the given point be at a or d, make $kl = at$ and $gn = bl$. Join al, dn. [Adding the quadrilateral ak to the congruent triangles with equal sides at, kl, we find al divides the trapezium in the required ratio. Then from VI. 1, dn does the same.]

As in Proposition 8, for any point t' between a and t, or t and d, we have a corresponding point k' between l and k or n and k, such that the line $t'k'$ divides the trapezium in the given ratio.

---

97 As 30, Favaro quotes, "Per rectam protractam super duo latera equidistantia quadrilaterum abscisum in data aliqua proportione dividere"; as 31: "Divisionem in eadem proportione ab angulis habere."

98 Here, as well as in 15 and 36, Leonardo introduces the representation of numbers by straight lines, and in considering these lines he invariably writes the word number in connection with them; *e. g.* 'number ez: number zi,' not $ez : zi$. Euclidean MSS. of the *Elements*, Books VII to IX, adopt this same method. In what follows, I shall use the abbreviated form.

If the given point be in *bg* at *l* or *n* or between *l* and *n*, the above reasoning suffices.

Suppose however that the given point were at *b*. Join *bd*. Through *n* draw *nc* parallel to *bd*. Join *bc*. Then *bc* divides the trapezium in the required ratio. Similarly for the point *g* and for any point between *b* and *n*, or between *g* and *l*.

Some of the parts which I have filled in above are covered by the general final statement: "nec non et diuidemus ipsum quadrilaterum ab omni puncto dato super aliquod laterum ipsius......" (Page 134, ll. 10–11. Compare Proposition 13.)

Proposition 10.

31. *"To divide a parallelogram into two equal parts by a straight line drawn from a given point outside the parallelogram."* [Leonardo 18, p. 124, ll. 5–7.]

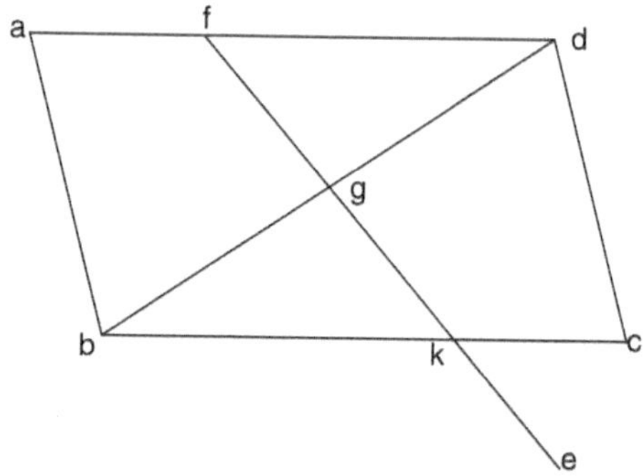

Let *abcd* be the given parallelogram and *e* the point outside. Join *bd* and bisect it in *g*. Join *eg* meeting *bc* in *k* and produce it to meet *ad* in *f*. Then the parallelogram has been divided into two equal parts by the line drawn through *e*, as may be proved by superposition; and one half is the quadrilateral *fabk*, the other, the quadrilateral *fkcd*[99].

[99] The proof also follows from the equality of the triangles *fgd*, *bgk*, by I. 26 and of the triangles *abd*, *bdc* by I. 4. This problem is possible for all positions of the point *e*.

PROPOSITION 11.

32. *"To cut off a certain fraction from a parallelogram by a straight line drawn from a given point outside of the parallelogram."*

This proposition is not explicitly formulated by Leonardo; but the general method he would have employed seems obvious from what has gone before.

Suppose it were required to cut off one-third of the given parallelogram ac by a line drawn through a point e outside of the parallelogram. Then by the method of Proposition 7, form a parallelogram two-thirds of ac. There are four such parallelograms with centres g_1, g_2, g_3, g_4. Lines l_1, l_2, l_3, l_4 through each one of these points and e will bisect a parallelogram (Proposition 10).

There are several cases to consider with regard to the position of e but it may be readily shown that, in one case at least, there is a line $l_i (i = 1, 2, 3, 4)$, which will cut off a third of the parallelogram ac.

Similarly for one-fourth, one-fifth, or any other fraction such as $m : n$ which represents the ratio of lengths of given lines.

PROPOSITION 12.

33. *"To divide a given trapezium into two equal parts by a straight line drawn from a point which is not situated on the longer side of the trapezium. It is necessary that the point be situated beyond the points of concourse of the two sides of the trapezium."* [Leonardo 28, p. 129, ll. 2–4, and another, unnumbered[100].]

PROPOSITION 13.

34. *"To cut off a certain fraction from a (parallel-) trapezium by a straight line which passes through a given point lying inside or outside the trapezium but so that a straight line can be drawn through it cutting both*

---

[100] As Leonardo 28 Favaro gives, "Qualiter quadrilatera duorum laterum equidistantium diuidi debeant a dato puncto extra figuram" and entirely ignores the paragraph headed, "De diuisione eiusdem generis, qua quadrilaterorum per rectam transeuntem per punctum datum infra ipsum" [p. 131, ll. 13–14].

the parallel sides of the trapezium[101]." [Part of Leonardo 32[102], p. 134, ll. 11–12.]

We first take up Leonardo's discussion of Proposition 12.

In the figure of Proposition 8, suppose al to be produced in the directions of the points e and r; tk in the directions of q and v, dn of z and h, cb of i and o, gf of s and p. Then for [any such exterior points e, q, z, i, s, r, v, h, o, p, lines are drawn bisecting the trapezium].

If the given point, x, were anywhere in the section of the plane above ad and between ea and dz, the line joining x to m would [by the same reasoning as in Proposition 8] bisect the trapezium. Similarly for all points below nl and between hn and lr [...... so also for all points within the triangles amd, nml].

This seems to be all that Euclid's Proposition 12 calls for. But just as Leonardo considers Proposition 8 for the general case with the given point any-

[101] The final clauses of Propositions 12 and 13, in Woepcke's rendering, are the same. I have given a literal translation in Proposition 12. Heath's translation and interpretation (after Woepcke) are given in 13. Concerning 12 and 13 Woepcke adds the following note: "Suppose it were required to cut off the nth part of the trapezium $ABDC$; make $A\alpha$ and $C\gamma$ respectively equal to the nth parts of AB and of CD; then $A\alpha\gamma C$ will be the nth part of the trapezium, for $\gamma\alpha$ produced will pass through the intersection of CA, DB produced. Now to draw through a given point E the transversal which cuts off a certain fraction of the trapezium, join the middle point μ of the segment $\alpha\gamma$, and the point E, by a line; this line EFG will be the transversal required to be drawn, since the triangle $\alpha F\mu$ equals the triangle $\gamma G\mu$.

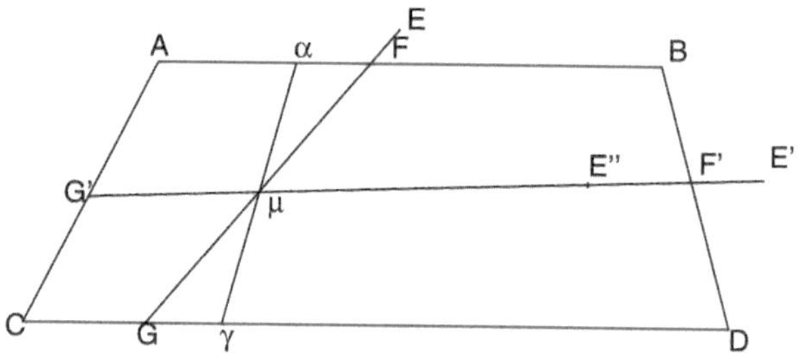

"But when the given point is situated as E' or E'' such that the transversal drawn through μ no longer meets the two parallel sides but one of the parallel sides and one of the two other sides, or the other two sides; then the construction indicated is not valid since $CG'\mu\gamma$ is not equal to $BF'\mu\alpha$. It appears that this is the idea which the text is intended to express. The 'points of concourse' are the vertices where a parallel side and one of the two other sides intersect; and the expression 'beyond' refers to the movement of the transversal represented as turning about the point μ."

[102] "Quadrilaterum [trapezium] ab omni puncto dato super aliquod laterum ipsius, et etiam ab omni puncto dato infra, uel extra diuidere in aliqua data proportioni."

where on the perimeter of the trapezium, so here, he discusses the constructions for drawing a line from any point inside or outside of a trapezium to divide it into two equal parts.

Leonardo does not give any details of the discussion of Euclid's Proposition 13, but after presentation of the cases given in Proposition 9 concludes: "et diuidemus ipsum quadrilaterum ab omni puncto dato super aliquod laterum ipsius, et etiam ab omni puncto dato infra uel extra" [Leonardo 32, p. 134, ll. 10–12].

From Leonardo's discussion in Propositions 8, 9, 12, not only are the necessary steps for the construction of 13 (indicated in the Woepcke note above[101]) evident, but also those for the more general cases, not considered by Euclid, where restrictions are not imposed on the position of the given point.

Proposition 14.

35. *"To divide a given quadrilateral into two equal parts by a straight line drawn from a given vertex of the quadrilateral."* [Leonardo 36, p. 138, ll. 10–11.]

Let *abcd* be the quadrilateral and *a* the given vertex. Draw the diagonal *bd*, meeting the diagonal *ac* in *e*. If *be*, *ed* are equal, [*ac* divides the quadrilateral as required].

If *be* be not equal to *ed*, make $bz = zd$.

Draw $zi \parallel ac$ to meet *dc* in *i*. Join *ai*. Then the quadrilateral *abcd* is divided as required by the line *ai*.

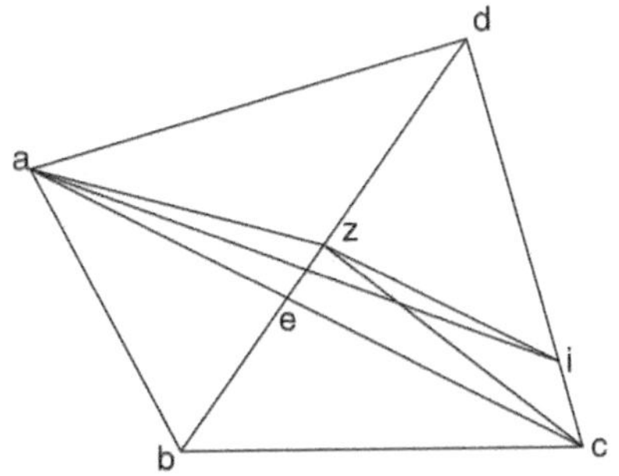

Proof: Join *az* and *zc*. Then the triangles *abz*, *azd* are respectively equal to the triangles *cbz*, *cdz*.

Therefore the quadrilateral *abcz* is one-half the quadrilateral *abcd*.

And since the triangles *azc*, *aic* are on the same base and between the same parallels *ac*, *zi*, they are equal.

To each add the triangle *abc*.

Then the quadrilateral *abcz* is equal to the quadrilateral *abci*. But the quadrilateral *abcz* is one-half of the quadrilateral *abcd*. Therefore *abci* is one-half of the quadrilateral *abcd*; "ut oportet."

Proposition 15.

36. *"To cut off a certain fraction from a given quadrilateral by a line drawn from a given vertex of the quadrilateral."* [Leonardo 40, p. 140, ll. 36–37.]

Let the given fraction be as $ez : zi$, and let the quadrilateral be *abcd* and the given vertex *d*. Divide *ac* in *t* such that

$$at : tc = ez : zi.$$

If *bd* pass through *t* [then *bd* is the line required].

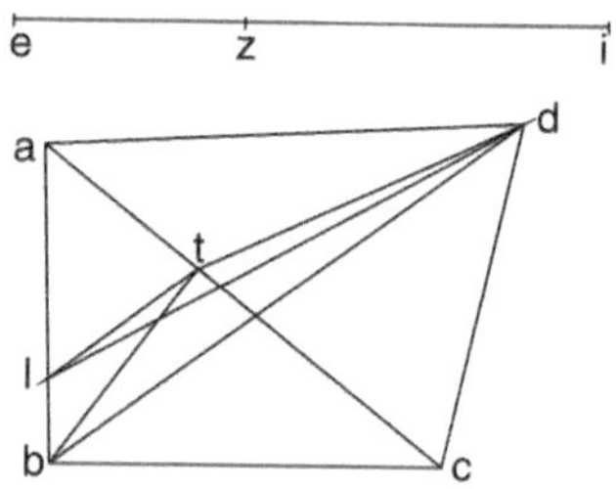

But if *bd* do not pass through *t* it will intersect either *ct* or *ta*; let it intersect *ct*. Join *bt*, *td*.

Then

$$\text{quadl.}tbcd : \text{quadl.}tbad = ct : ta = ez : zi.$$

Draw *tl* parallel to the diagonal *bd*, and join *dl*. Then the quadrilaterals *lbcd*, *tbcd* are equal and the construction has been made as required; for

$$ct : ta = ez : zi = \text{quadl.}lbcd : \triangle dal.$$

And if *bd* intersect *ta* [a similar construction may be given to divide the given quadrilateral, by a line through *d*, into a quadrilateral and triangle in the required ratio].

Leonardo then gives the construction for dividing a quadrilateral in a given ratio by a line drawn through a point which divides a side of the quadrilateral in the given ratio.

PROPOSITION 16.

37. *"To divide a given quadrilateral into two equal parts by a straight line drawn from a given point situated on one of the sides of the quadrilateral."* [Leonardo 37, p. 138, ll. 28–29.]

Let *abcd* be the given quadrilateral, *e* the given point. Divide *ac* into two equal parts by the line *dt* [Prop. 14]. Join *et*. The line *et* either is, or is not, parallel to *dc*.

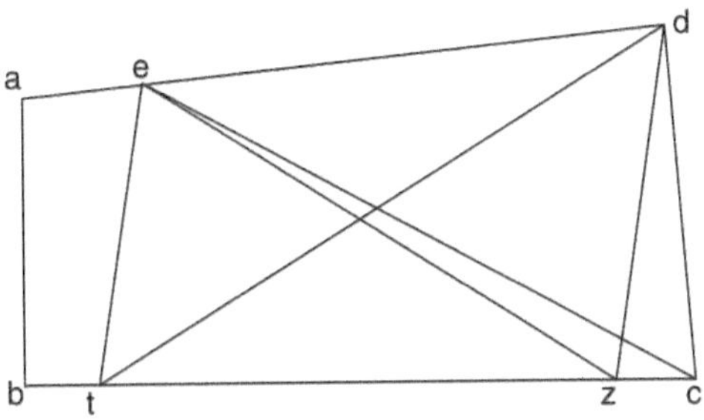

(Two of Leonardo's figures are combined in one, here.)

If *et* be parallel to *dc*, join *ec*. Then the quadrilateral *ac* [is bisected by the line *ec*, etc.].

If *et* be not parallel to *dc*, draw $dz \parallel et$. Join *ez*. Then *ac* [is bisected by the line *ez*, etc.].

Leonardo does not consider the case of failure of this construction, namely when *dz* falls outside the quadrilateral. Suppose in such a case that the problem were solved by a line joining *e* to a point z' (not shown in the figure) on *dc*. Through *t*, draw $tt' \parallel cd$. Join ct'. Then $\triangle ct'd = \triangle ctd = \triangle edz'$. Whence $\triangle et'c = \triangle ez'c$, or $t'z' \parallel ce$. Therefore from t', z' may be found and the solution in this case is also possible, indeed in more than one way, but it is not in Euclid's manner to consider this question.

Should the diagonal *db* bisect the quadrilateral *ac*, the discussion is similar to the above.

But if the line drawn from *d* to bisect the quadrilateral meet the side *ab* in *i*, draw *bk* bisecting the quadrilateral *ac*.

If *k* be not the given point, it will be between *k* and *d* or between *k* and *a*.

In the first case join *be* and through *k* draw $kl \parallel eb$. Join *el* [then *el* is the required bisector].

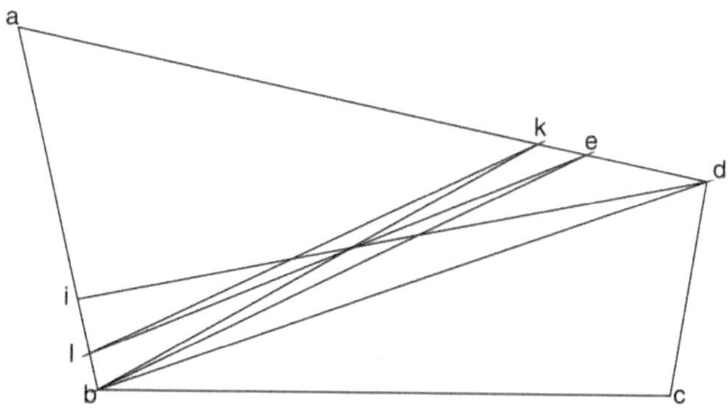

If the point e be between a and k [a similar construction with the line through k parallel to be, and meeting bc in m, leads to the solution by the line em].

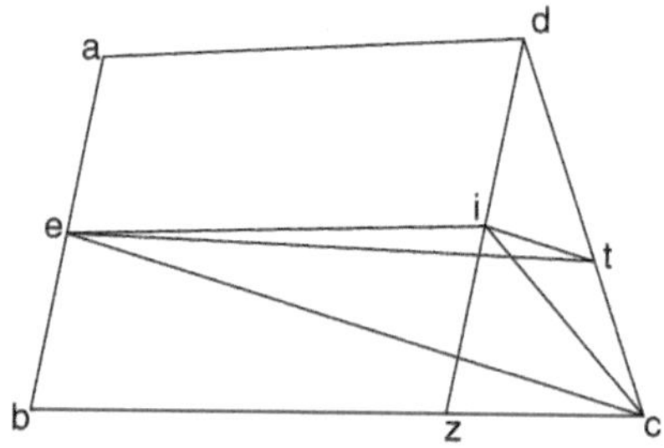

Were e at the middle of a side such as ab, draw $dz \parallel ab$ and bisect dz in i. Join ei, ci and ec. Through i draw $it \parallel ec$. Join et; then et [bisects the quadrilateral ac, since $\triangle itc = \triangle ite$, etc.].

If dz were to fall outside the quadrilateral, draw from c the parallel to ba; and so on.

Proposition 17.

38. *"To cut off a certain fraction from a quadrilateral by a straight line drawn from a given point situated on one of the sides of the quadrilateral."* [Leonardo 39, p. 140, ll. 11–12.]

Let *abcd* be the given quadrilateral and suppose it be required to cut off one-third by a line drawn from the point *e* in the side *ad*.

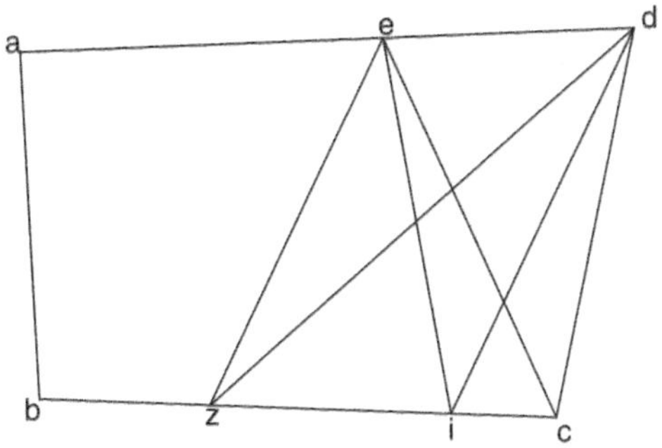

Draw *dz* cutting off one-third of *ac* [Prop. 15].

Join *ez*, *ec*.

If $ez \parallel dc$, then *ecd* [is the required part cut off, etc.].

But if *ez* be not parallel to *dc*, draw $di \parallel ez$ and join *ei*. [Then this is the line required, etc.]

The case when *ei* cuts *dc* is not taken up but it may be considered as in the last proposition.

So also to divide *ac* into any ratio: draw *dz* dividing it in that ratio (Prop. 15), and then proceed as above.

A particular case which Leonardo gives may be added.

Let *ab* be divided into three equal parts *ae*, *ef*, *fb*; draw $dg \parallel ab$ and cut off $gh = \frac{1}{3}gd$. Join *fc* and through *h* draw $hi \parallel fc$, meeting *dc* in *i*. Join *fi*; and the quadrilateral *fbci* will be one-third of the quadrilateral *ac*. [As in latter part of Prop. 16.]

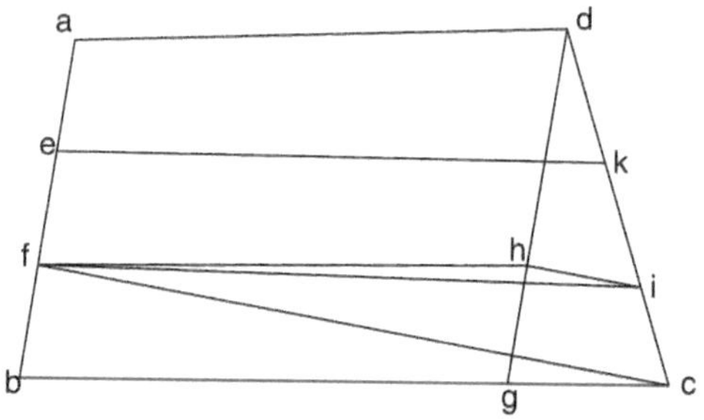

Then *ek* may be drawn to bisect the quadrilateral *afid* [Prop. 16], and thus the quadrilateral *abcd* will be divided into three equal portions which are the quadrilaterals *ak*, *ei*, *fc*.

PROPOSITION 18.

39. *"To apply to a straight line a rectangle equal to the rectangle contained by AB, AC and deficient by a square*[103]*."*

---

[103] This proposition is interesting as illustrating the method of *application of areas* which was "one of the most powerful methods on which Greek Geometry relied." The method first appears in the *Elements* in I. 44: *To a given straight line to apply, in a given rectilineal angle, a parallelogram equal to a given triangle*—a proposition which Heath characterises as "one of the most impressive in all geometry" while the "marvellous ingenuity of the solution is indeed worthy of the 'godlike men of old' as Proclus calls the discoverers of the method of 'application of areas'; and there would seem to be no reason to doubt that the particular solution, like the whole theory, was Pythagorean, and not a new solution due to Euclid himself."

[I continue to quote mainly from Heath who may be consulted for much greater detail: HEATH, *Thirteen Books of Euclid's Elements*, I, 9, 36, 343–7, 383–8; II, 187, 257–67—HEATH, *Apollonius of Perga Treatise on Conic Sections*, Cambridge, 1896, pp. lxxxi–lxxxiv, cii–cxi—HEATH, *The Works of Archimedes*, Cambridge, 1897, pp. xl–xlii, 110 and "Equilibrium of Planes," Bk II, Prop. 1, and "On conoids and spheroids," Props. 2, 25, 26, 29. See also: CANTOR, *Vorlesungen über Geschichte der Math.* I$_3$, 289–291, etc. (under index heading 'Flächenanlegung')—H. G. ZEUTHEN, *Geschichte der Mathematik im Alterthum und Mittelalter*, Kopenhagen, 1896, pp. 45–52 (French ed. Paris, 1902, pp. 36–44)— C. TAYLOR, *Geometry of Conics...*, Cambridge, 1881, pp. XLIII–XLIV.]

The simple *application* of a parallelogram of given area to a given straight line as one of its sides is what we have in the *Elements* I. 44 and 45; the general form of the problem with regard to *exceeding* and *falling-short* may be stated thus:

"To apply to a given straight line a rectangle (or, more generally, a parallelogram) equal to a given rectilineal figure and (1) *exceeding* or (2) *falling-short* by a square (or, in the more general case, a parallelogram similar to a given parallelogram)."

What is meant by saying that the applied parallelogram (1) *exceeds* or (2) *falls short* is that, while its base coincides and is coterminous *at one end* with the straight line,

"After having done what was required, if some one ask, How is it possible to apply to the line AB a rectangle such that the rectangle $AE \cdot EB$

is equal to the rectangle $AB \cdot AC$ and deficient by a square—we say that it is impossible, because AB is greater than BE and AC greater than AE, and consequently the rectangle $BA \cdot AC$ greater than the rectangle $AE \cdot EB$. Then when one applies to the line AB a parallelogram equal to the rectangle $AB \cdot AC$ the rectangle $AZ \cdot ZB$ is......[104]."

In this problem it is required to find in the given line AB a point Z such that

$$AB \cdot ZB - ZB^2 [= AZ \cdot ZB \text{ by II. 3; cf. X. 16 lemma}] = AB \cdot AC.$$ [105]

---

the said base (1) overlaps or (2) falls short of the straight line *at the other end*, and the portion by which the applied parallelogram exceeds a parallelogram of the same angle and height on the given straight line (exactly) as base is a parallelogram similar to a given parallelogram (or, in particular cases, a square). In the case where the parallelogram is to *fall short*, some such remark as Woepcke's (note 104) is necessary to express the condition of possibility of solution. For the other case see note 116.

The solution of the problems here stated is equivalent to the solution of a quadratic equation. By means of II. 5 and 6 we can solve the equations

$$ax \pm x^2 = b^2,$$
$$x^2 - ax = b^2,$$

but in VI. 28, 29 Euclid gives the equivalent of the solution of the general equations

$$ax \pm px^2 = A.$$

VI. 28 is: *To a given straight line to apply a parallelogram equal to a given rectilineal figure and deficient by a parallelogrammic figure similar to a given one: thus the given rectilineal figure must not be greater than the parallelogram described on the half of the straight line and similar to the defect.*

The Proposition 18 of Euclid under consideration is a particular case of this problem and as the fragment of the text and Woepcke's note (note 104) are contained in it, doubt may well be entertained as to whether Euclid gave any construction in his book *On Divisions.* The problem can be solved without the aid of Book VI of the *Elements* and by means of II. 5 and II. 14 only, as indicated in the text above.

The appropriation of the terms parabola (*application*), hyperbola (*exceeding*) and ellipse (*falling-short*) to conic sections was first introduced by Apollonius as expressing in each case the fundamental properties of curves as stated by him. This fundamental property is the geometrical equivalent of the Cartesian equation referred to any diameter of the conic and the tangent at its extremity as (in general, oblique) axes. More particulars in this connection are given by Heath.

The terms "parabolic," "hyperbolic" and "elliptic," introduced by Klein for the three main divisions of Geometry, are appropriate to systems in which a straight angle equals, exceeds and falls short of the angle sum of any triangle. *Cf.* W. B. FRANKLAND, *The First Book of Euclid's Elements with a Commentary based principally upon that of Proclus Diadochus*... Cambridge, 1905, p. 122.

Find, by II. 14, the side, b, of a square equal in area to the rectangle $AB.AC$, then the problem is exactly equivalent to that of which a simple solution was given by Simson[106]:

To apply a rectangle which shall be equal to a given square, to a given straight line, deficient by a square: but the given square must not be greater than that upon the half of the given line.

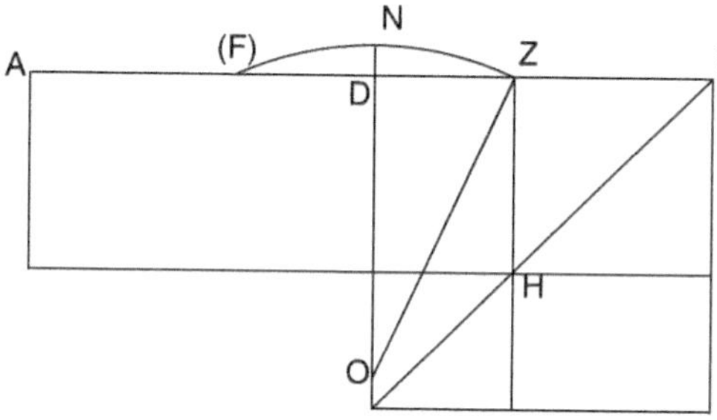

Bisect AB in D, and if the square on AD be equal to the square on b, the thing required is done. But if it be not equal to it, AD must be greater than b according to the determination. Then draw DO perpendicular to AB and equal to b; produce OD to N so that $ON = DB$ (or $\frac{1}{2}a$); and with O as centre and radius ON describe a circle cutting DB in Z.

Then ZB (or x) is found, and therefore the required rectangle AH.

For the rectangle $AZ . ZB$ together with the square on DZ is equal to the square on DB, [II. 5]

i. e. to the square on OZ,

i. e. to the squares on OD, DZ. [I. 47]

Whence the rectangle $AZ . ZB$ is equal to the square on OD.

---

104 Woepcke here remarks: "Evidently if a denote the length of the line to which the rectangle is to be applied, Problem 18 is only possible when

$$AB . AC < \left(\frac{a}{2}\right)^2 .$$

Then if a be taken as AB one of the two sides of the given rectangle, relatively to the other side, $AC < \frac{AB}{4}$. It is probably the demonstration of this which was given in the missing portion of the text."

105 If $AB = a$, $ZB = x$, $AB . AC = b^2$, the problem is to find a geometric solution of the equation $ax - x^2 = b^2$. Ofterdinger[38] (p. 15) seems to have quite missed the meaning of this problem. He thought, apparently, that it was equivalent to X. 16, lemma, of the *Elements*.

106 R. SIMSON, *Elements of Euclid*, ninth ed., Edinburgh, 1793, pp. 335–6.

Wherefore the rectangle AH equals the given square upon b (i. e. the rectangle $AB \cdot AC$) and has been applied to the given straight line AB, deficient by the square HB[107].

Proposition 19.

40. *"To divide a given triangle into two equal parts by a line which passes through a point situated in the interior of the triangle."* [Leonardo 3, p. 115, ll. 7–10.]

"Let the given triangle be ABC, and the given point in the interior of this triangle, D.

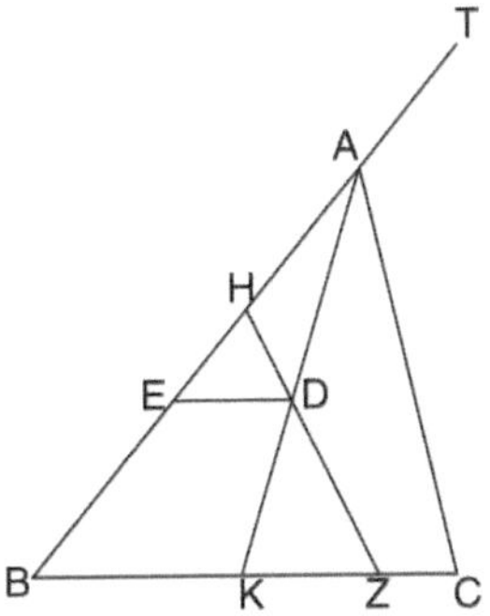

It is required to draw through D a straight line which divides the triangle ABC into two equal parts.

Draw from the point D a line parallel to the line BC, as DE, and

Apply to DE a rectangle equal to half of the rectangle $AB \cdot BC$, such as

$$TB \cdot DE \left[TB = \frac{AB \cdot BC}{2DE} \right].$$

Apply to the line TB a parallelogram equal to the rectangle $BT \cdot BE$ and deficient by a square[1071]. [Prop. 18]

---

[107] It is not in the manner of Euclid to take account of the two solutions found by considering (F), as well as Z, determined by the circle with centre O.

Although Leonardo's construction for Problem 19 is identical with that of Euclid who makes use of Problem 18, Leonardo does not seem to have anywhere formulated Problem 18. He may have considered it sufficiently obvious from VI. 28, or from II. 5 and II. 6, of which he gives the enunciations in the early pages (15–16) of his *Practica Geometriae*; he also considers (p. 60) the roots of a resulting quadratic equation, $ax - x^2 = b$ (*cf.* CANTOR, *Vorlesungen...*, II_2, 39), but does not give II. 14. *Cf. Bibliotheca Mathematica*, (3), 1907–8, VIII, 190; and also IX, 245.

Let the rectangle applied be

$$BH \cdot HT[(TB - HT) \cdot HT = TB \cdot BE].$$

Draw the line HD and produce it to Z.

Then this is the line required and the triangle ABC is divided into two equal parts HBZ and $HZCA$.

Demonstration. The rectangle $TB.BE$ is equal to the rectangle $TH.HB$, whence it follows that

$$BT : TH = HB : BE;$$

then *dividendo*[108] $TB : BH = BH : HE.$

But $BH : HE = BZ : ED;$ [VI. 2]

therefore $TB : BH = BZ : ED.$

Consequently the rectangle $TB \cdot ED$ is equal to the rectangle $BH \cdot BZ$. But the rectangle $TB \cdot ED$ is equal to half the rectangle $AB \cdot BC$; and

$$BH \cdot BZ : AB \cdot BC = \triangle HBZ : \triangle ABC^{88},$$

since the angle B is common. The triangle HBZ is, then, half the triangle ABC.

Therefore the triangle ABC is divided into two equal parts BHZ and $AHZC$.

If, in applying to TB a parallelogram equal to the rectangle $TB.BE$ and of which the complement is a square, we obtain the rectangle $AB \cdot AT$[109],

---

[1071] The corresponding sentence in Leonardo is (p. 115, ll. 15–17): "Deinde linee *gz* applicabis paralilogramum deficiens figura tetragona, quod sit equale superficies *ge* in *gz*."

[108] "Elements, Book V, definition 16" (Woepcke). This is definition 15 in HEATH, *The Thirteen Books of Euclid's Elements*, II, 135.

[109] "In other words when H coincides with A. This can only be the case when D is situated on the line which joins A to the middle of the base BC" (Woepcke). If D were at the centre of gravity of the triangle, three lines could be drawn through D dividing the triangle into two equal parts. As introductory to his Prop. 3, Leonardo proved that the medians of a triangle meet in a point, and trisect one another—results known to Archimedes[601], but no complete, strictly geometric proof has come down to us from the Greeks. Leonardo then proves that if a point be taken on any one of the medians, or on one of the medians produced, the line through this point and the corresponding angular point of the triangle will divide the triangle into two equal parts. He next shows that lines through the vertices of a triangle and any point within not on one of the medians, will divide the given triangle into triangles whose areas are each either greater than or less than the area of half of the original triangle. This leads Leonardo to the consideration of the problem, to draw through a point, within a triangle and not on one of the medians, a line which *will* bisect the area of the triangle. (Euclid, Prop. 19.)

The last paragraph of Euclid's proof, as it has come down to us through Arabian sources, does not ring true, and it was not in the Euclidean manner to consider special cases.

After Leonardo's proof of Proposition 19, a numerical example is given.

we may demonstrate in an analogous manner, by drawing the line AD and prolonging it to K, that the triangle ABK is one-half of the triangle ABC. And this is what was required to be demonstrated."

Proposition 20.

41. *"To cut off a certain fraction from a given triangle by a line drawn from a given point situated in the interior of the triangle."* [Leonardo 10, p. 121, ll. 1–2.]

"Let ABC be the given triangle and D the given point in the interior of the triangle. It is required to pass through the point D a straight line which cuts off a certain fraction of the triangle ABC.

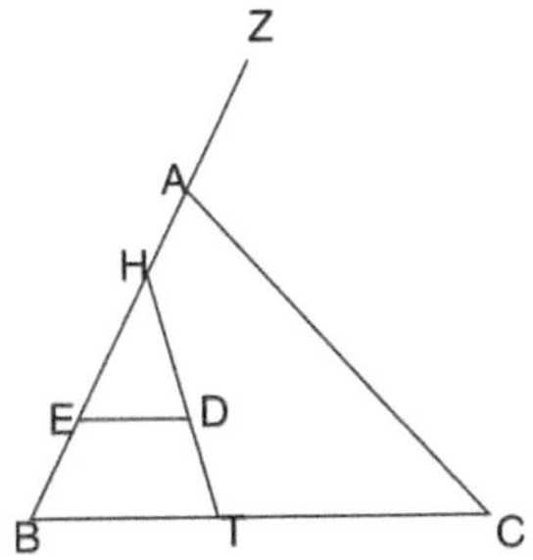

"Let the certain fraction be one-third. Draw from the point D a line parallel to the line BC, as DE, and apply to DE a rectangle equal to one-third of the rectangle $AB \cdot BC$. Let this be

$$BZ \cdot ED \left[BZ = \frac{AB \cdot BC}{3 \cdot ED} \right].$$

Then apply to ZB a rectangle equal to the rectangle $ZB \cdot BE$ and deficient by a square. [Prop. 18.] Let the rectangle applied be the rectangle

$$BH \cdot HZ \, [(ZB - HZ)\, HZ = ZB \cdot BE].$$

Draw the line HD and produce it to T.

"On proceeding as above we may demonstrate that the triangle HTB is one-third of the triangle ABC; and by means of an analogous construction to this we may divide the triangle in any ratio. But this is what it is required to do[110]."

[110] Leonardo gives the details of the proof for the case of one-third and does not refer

PROPOSITION 21.

42. *"Given the four lines A, B, C, D and that the product of A and D is greater than the product of B and C; I say that the ratio of A to B will be greater than the ratio of C to D"*[111].

---

to any other fraction. If, however, the "certain fraction" were the ratio of the lengths of two given lines, $m : n$, we could readily construct a rectangle equal to $\frac{m}{n} \cdot AB \cdot BC$, and then find the rectangle $BZ \cdot ED$ equal to it. The rest of the construction is the same as given above.

According to the conditions set forth in Proposition 18, there will be two, one, or no solutions of Propositions 19 and 20. Leonardo considers only the Euclidean cases. *Cf.* notes 104 and 107.

The case where there is no solution may be readily indicated. Suppose, in the above figure, that $BE = EH$, then of all triangles formed by lines drawn through D to meet AB and BC, the triangle HBT has the minimum area. (Easily shown synthetically as in D. CRESSWELL, *An Elementary Treatise on the Geometrical and Algebraical Investigations of Maxima and Minima.* Second edition, Cambridge, 1817, pp. 15–17.) Similar minimum triangles may be found in connection with the pairs of sides AB, AC and AC, CB. Suppose that neither of these triangles is less than the triangle HBT. Then if

$$\triangle HBT : \triangle ABC > m : n,$$

the solution of the problem is impossible.

[111] This and the next four auxiliary propositions for which I supply possible proofs, seem to be neither formally stated nor proved by Leonardo. At least some of the results are nevertheless assumed in his discussion of Euclid's later propositions, as we shall presently see. Although these auxiliary propositions are not given in the *Elements*, they are assumed as known by Archimedes, Ptolemy and Apollonius.

For example, in Archimedes' "On Sphere and Cylinder," II. 9 (Heiberg, ed. I, 1910, p. 227; Heath, ed. 1897, p. 90), Woepcke 21 is used. See also Eutocius' Commentary (Archimedis *Opera omnia* ed. Heiberg, III, 1881, p. 257, etc.), and HEIBERG, *Quaestiones Archimedeae*, Hauniae, 1879, p. 45 f. For a possible application by Archimedes (in his *Measurement of a circle*) of what is practically equivalent to Woepcke 24, see Heath's *Archimedes...*, 1897, p. xc.

The equivalent of Woepcke 24 is assumed in the proof of a proposition given by Ptolemy (87–165 A. D.) in his *Syntaxis*, vol. I, Heiberg edition, Leipzig, 1898, pp. 43–44. This in turn is tacitly assumed by Aristarchus of Samos (*circa* 310–230 B. C.) in his work *On the Sizes and Distances of the Sun and Moon* (see Heath's edition *Aristarchus of Samos the Ancient Copernicus*, Oxford, 1913, pp. 367, 369, 377, 381, 389, 391).

As to the use of the auxiliary propositions in the two works *Proportional Section* and *On Cutting off a Space*, of Apollonius, we must refer to Pappus' account (Pappi Alexandrini *Collectionis...* ed. Hultsch, vol. II, 1877, pp. 684 ff.). Woepcke 21, 22 occur on pp. 696–697; Woepcke 24 enters on pp. 684–687; Woepcke 23, 25 are given on pp. 687, 689. Perhaps this last statement should be modified; for whereas Euclid's propositions affirm that if

$$a : b \gtreqless c : d, \quad a - b : b \gtreqless c - d : d,$$

Pappus shows that if

$$a : b \gtreqless c : d, \quad a : a - b \lesseqgtr c : c - d;$$

Given $A \cdot D > B \cdot C$. To prove $A : B > C : D$.

---

but these propositions are immediately followed by others which state that if

$$a : b \gtrless c : d, \text{ then } b : a \lessgtr d : c.$$

Below is given a list of the various restorations of the above-named works of Apollonius, based on the account of Pappus. By reference to these restorations the way in which the auxiliary propositions are used or avoided may be observed. We have already (Art. 21) noticed a connection of Apollonius' work *On Cutting off a Space* with our subject under discussion. Some of these titles will therefore supplement the list given in the Appendix.

Wilebrordi Snellii R. F. περὶ λόγου ἀποτομὲς καὶ περὶ ςηωρίου ἀποτομης (Apollonii) resuscitata geometria. Lugodini, *ex officina Platiniana* Raphelengii, MD.CVII pp. 23.

More or less extensive abridgment of Snellius's work is given in:

(a) *Universae geometriae mixtaeque mathematicae synopsis et bini refractionum demonstratarum tractatus. Studio et opera F. M. Mersenni.* Parisiis, M.DC.XLIV, p. 382.

(b) *Cursus mathematicus*, P. Herigone. Paris, 1634, tome I, pp. 899–904; also Paris, 1644.

Apollonii Pergaei de sectione rationis libri duo ex Arabico MS[to] *Latine versi accedunt ejusdem de sectione spatii libri duo restituti... opera & studio Edmundi Halley...* Oxonii,... MDCCVI, pp. 8 + liii + 168.

(a) *Die Bücher des Apollonius von Perga De sectione rationis nach dem Lateinischen des Edm. Halley frey bearbeitet, und mit einem Anhange versehen von W. A. Diesterweg*, Berlin, 1824, pp. xvi + 218 + 9 pl.

(b) *Des Apollonius von Perga zwei Bücher vom Verhältnissschnitt* (*de sectione rationis*) *aus dem Lateinischen des Halley übersetzt und mit Anmerkungen begleitet und einem Anhang versehen von August Richter* ... Elbing, 1836, pp. xxii + 143 + 4 pl.

Die Bücher des Apollonius von Perga de sectione spatii wiederhergestellt von Dr W. A. Diesterweg... Elberfeld, 1827... pp. vi + 154 + 5 pl.

Des Apollonius von Perga zwei Bücher vom Raumschnitt. Ein Versuch in der alten Geometrie von A. Richter. Halberstadt, 1828, pp. xvi + 105 + 9 pl.

Die Bücher des Apollonius von Perga de sectione spatii, analytisch bearbeitet und mit einem Anhange von mehreren Aufgaben ähnlicher Art versehen von M. G. Grabow... Frankfurt a. M., 1834, pp. 80 + 3 pl.

Geometrische Analysis enthaltend des Apollonius von Perga sectio rationis, spatii und determinata, nebst einem Anhange zu der letzten, neu bearbeitet vom Prof. Dr Georg Paucker, Leipzig, 1837, pp. xii + 167 + 9 pl.

M. Chasles discovered that by means of the theory of involution a single method of solution could be applied to the main problem of the three books of Apollonius above mentioned. This solution was first published in *The Mathematician*, vol. III, Nov. 1848, pp. 201–202. This is reproduced by A. Wiegand in his *Die schwierigeren geometrischen Aufgaben aus des Herrn Prof. C. A. Jacobi Anhängen zu Van Swinden's Elementen der Geometrie. Mit Ergänzungen englischer Mathematiker...* Halle, 1849, pp. 148–149, and it appears at greater length in Chasles' *Traité de Géométrie supérieure*, Paris,

Let the lines A, D be adjacent sides of a rectangle; and let there be another rectangle with side B lying along A and side C along D. Then either A is greater than B, or D greater than C, for otherwise the rectangle $A \cdot D$ would not be greater than the rectangle $B \cdot C$.

Let then $A > B$. To D apply the rectangle $B.C$ and we get a rectangle

$$A' \cdot D = B \cdot C; \qquad [\text{I. } 44\text{–}45]$$

then

$$A' : B = C : D. \qquad [\text{VII. } 19]$$

But since

$$A > A',$$

$$A : B > A' : B; \qquad [\text{V. } 8]$$

$$\therefore A : B > C : D. \qquad [\text{V. } 13]$$

Q. E. D.

Pappus remarks: Conversely if $A : B > C : D$, $A \cdot D > B \cdot C$. The proof follows at once.

For, find A' such that $A' : B = C : D$;

then $A : B > A' : B$,

and $A > A'$. But $A' \cdot D = B \cdot C$. $\therefore A \cdot D > B \cdot C$. Q. E. D.

Proposition 22.

43. *"And when the product of A and D is less than the product of B and C, then the ratio of A to B is less than the ratio of C to D."*

From the above proof we evidently have

$$C : D > A : B,$$

that is,

$$A : B < C : D.$$

Conversely, as above, if $A : B < C : D$, $A \cdot B < C \cdot D$.

1852, pp. 216–218; 2e éd. 1880, pp. 202–204. It was no doubt Chasles who inspired *Die Elemente der projectivischen Geometrie in synthetischer Behandlung. Vorlesungen von H. Hankel*, (Leipzig, 1875), "Vierter Abschnitt, Aufgaben des Apollonius," pp. 128–145; "sectio rationis," pp. 128–138; "sectio spatii," pp. 138–140.

The *"Three Sections," the "Tangencies" and a "Loci Problem" of Apollonius... by M. Gardiner*, Melbourne, 1860. Reprinted from the *Transactions of the Royal Society of Victoria*, 1860–1861, V, 19–91 + 10 pl.

Die sectio rationis, sectio spatii und sectio determinata des Apollonius nebst einigen verwandten geometrischen Aufgaben von Fr. von Lühmann. Progr. Königsberg in d. N. 1882, pp. 16 + 1 pl.

"Ueber die fünf Aufgaben des Apollonius," *von L. F. Ofterdinger. Jahreshefte des Vereines für Math. u. Naturwiss. in Ulm a. D.* 1888, I, 21–38; "Verhältnissschnitt," pp. 23–25; "Flächenschnitt," pp. 26–27.

It is really this converse, and not the proposition, which Euclid uses in Proposition 26. Proclus remarks (page 407) that the converses of Euclid's Elements, I. 35, 36, about parallelograms, are unnecessary "because it is easy to see that the method would be the same, and therefore the reader may properly be left to prove them for himself." No doubt similar comment is justifiable here.

PROPOSITION 23.

44. *"Given any two straight lines and on these lines the points A, B, and D, E; and let the ratio of AB to BC be greater than the ratio of DE : EZ; I say that* dividendo *the ratio of AC to CB will be greater than the ratio of DZ to ZE."*

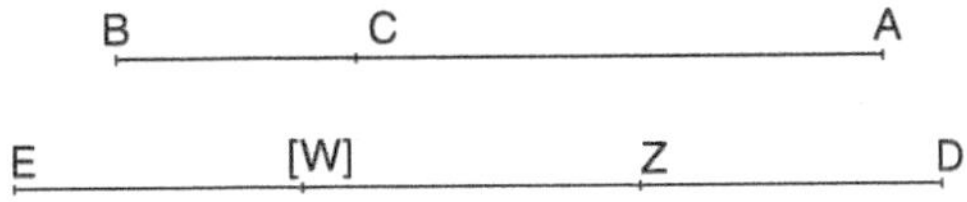

Given $AB : BC > DE : EZ.$

To prove $AC : CB > DZ : ZE.$

To AB, BC, DE find a fourth proportional EW. [VI. 12]

Then $AB : BC = DE : EW.$ ....(1)

But $AB : BC > DE : EZ;$

$\therefore DE : EW > DE : EZ.$ [V. 13]

$\therefore EW < EZ.$ [V. 8]

From (1) $AC : CB = DW : WE;$ ....(2) [V. 17]

since $DW > DZ, \quad DW : WE > DZ : WE.$ [V. 8]

$\therefore AC : CB > DZ : WE.$ [V. 13]

But $WE < ZE; \quad \therefore DZ : WE > DZ : ZE.$ [V. 8]

$\therefore AC : CB > DZ : ZE.$ From (2) and [V. 17]

Q. E. D.

PROPOSITION 24.

45. *"And in an exactly analogous manner I say that when the ratio of AC to CB is greater than the ratio of DZ to ZE, we shall have* componendo[112] *the ratio of AB to BC is greater than the ratio of DE to EZ."*

[112] "Elements, Book V, definition 15" (Woepcke). This is definition 14 in HEATH, *The Thirteen Books of Euclid's Elements*, II, 135.

Given $AC : CB > DZ : ZE.$

To prove $AB : BC > DE : EZ.$

Determine W, as before, such that

$$AB : BC = DE : EW.$$

Then $AC : CB = DW : WE.$ [V. 17]

$\therefore DW : WE > DZ : ZE.$ ......(1) [V. 13]

Now either $EW > EZ$ or $EW < EZ.$

If $EW > EZ$, $DW < DZ$, and

$$DW : EW < DZ : EW.$$ [V. 8]

So much the more is

$$DW : EW < DZ : EZ$$ [V. 8]

which contradicts (1).

$$\therefore EW < EZ.$$

But $AB : BC = DE : EW,$

and $DE : EW > DE : EZ;$ [V. 8]

$\therefore AB : BC > DE : EZ.$ [V. 13]

Q. E. D.

PROPOSITION 25.

46. *"Suppose again that the ratio of AB to BC were less than the*

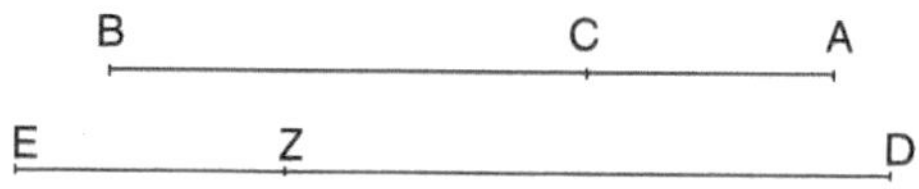

ratio of DE to EZ; dividendo *the ratio of AC to CB will be less than the ratio of DE to ZE*[113]*."*

Just as the proof of Proposition 22 was contained in that for Proposition 21, so here, the proof required is contained in the proof of Proposition 23. Similarly the converse of Proposition 25 flows out of 24.

---

[113] The auxiliary propositions are introduced, apparently, to assist in rendering, with faultless logic, the remarkable proof of Proposition 26. In this proof it will be observed that we are referred back to Proposition 21, to the converse of Proposition 22 and to Proposition 25 only, although 23 is really the same as 25. But no step in the reasoning has led to Proposition 24. If this is unnecessary, why has it been introduced?

PROPOSITION 26.

47. *"To divide a given triangle into two equal parts by a line drawn from a given point situated outside the triangle."* [Leonardo 4, p. 116, ll. 35–36.]

Let the triangle be abg and d the point outside.

Join ad and let ad meet bg in e. If $be = eg$, what was required is done. For the triangles abe, aeg being on equal bases and of the same altitude are equal in area.

But if be be not equal to eg, let it be greater, and draw through d, parallel to bg, a line meeting ab produced in z.

Since
$$be > \tfrac{1}{2} bg,$$
$$\text{area } ab \cdot be > \tfrac{1}{2} \text{ area } ab \cdot bg; \qquad [Cf. \text{ VII. } 17]$$
much more then is
$$\text{area } ab \cdot zd > \tfrac{1}{2} \text{ area } ab \cdot bg, \qquad \text{since } zd > be.$$

Now take
$$\text{area } ib \cdot zd = \tfrac{1}{2} \text{ area } ab \cdot bg; \qquad [\text{I. } 44]$$
then
$$\text{area } ab \cdot be > \text{ area } ib \cdot zd,$$
and
$$zd : be < ba : bi.\qquad [\text{Prop. 21 or 22}]$$

[114]

But
$$zd : be = za : ab, \qquad [\text{VI. } 4]$$
$$\therefore zb : ba < ai : ib; \qquad [\text{V. 13 and Prop. 25}]$$
or
$$\text{area } zb \cdot bi < \text{area } ba \cdot ai. \qquad [\text{Converse of Prop. 22}]$$

To answer this question, let us inspect the auxiliary propositions more closely. In a sense Propositions 21 and 22 go together: If $ad \gtreqless bc$, then $a : b \gtreqless c : d$. So also for Propositions 23 and 25: If $a : b \gtreqless c : d$, then $a - b : b \gtreqless c - d : d$. Proposition 24 is really the converse of 23: If $a : b > c : d$, then $a + b : b > c + d : d$. Had Euclid given another proposition: If $a : b < c : d$ then $a + b : b < c + d : d$, we should have had two groups of propositions 21, 22, and 23, 25 with their converses. Now the converses of 21 and 22 are exceedingly evident in both statement and proof. But this can hardly be said of the proof of 24, the converse of 23. The converse of 23 having been given the formulation of the statement and proof of the converse of 25 is obvious and unnecessary to state, according to Euclid's ideals (cf. Art. 43). It might therefore seem that Proposition 24 is merely given to complete what is not altogether obvious, in connection with the statement of the four propositions 21 and 22, 23 and 25, and their converses. In Pappus' discussion some support is given to this view, since Propositions 21 and 22 and converses are treated as a single proposition; Propositions 23, 25 as another proposition, while the converses of 23 and 25 are dealt with separately.

The more probable explanation is, however, that Propositions 23 and 24 were given by Euclid because they were necessary for the discussion of other cases of Proposition 26 (assuming that the first case of Leonardo was that given by Euclid), for it was not his manner to consider different cases. Indeed if we take be less than ge in the first part of Leonardo's discussion exactly Propositions 23 and 24 are necessary.

Apply a rectangle equal to the rectangle $zb \cdot bi$ to the line bi, but exceeding by a square[115]; that is to bi apply a line such that when multiplied by itself and by bi the sum will be equal to the product of zb and bi; let ti be the side of the square[116].

---

[114] Therefore $bi < ba$, and if bi be measured along ba, i will fall between b and a.

[115] Here again we have an expression with the true Greek ring: "adiungatur quidem recte *.bi.* paralilogramum superhabundans figura tetragona equale superficiei *.zb.* in *.bi.*"

[116] We we have seen that i lies between b and a. And since it has been shown that $zb \cdot bi < ba \cdot ai$, we now have $ba \cdot ai > bt \cdot ti$. If $bt > ba$, ti is also greater than ai, and $bt \cdot ti \not< ba \cdot ai$. Therefore $bt < ba$ and t falls between b and a. But it also falls between a and i by reason of the construction (always possible) which is called for.

In his book on *Divisions* (*of figures*) Euclid does not formulate the proposition here quoted, possibly because of its similarity to Proposition 18 (see note 103).

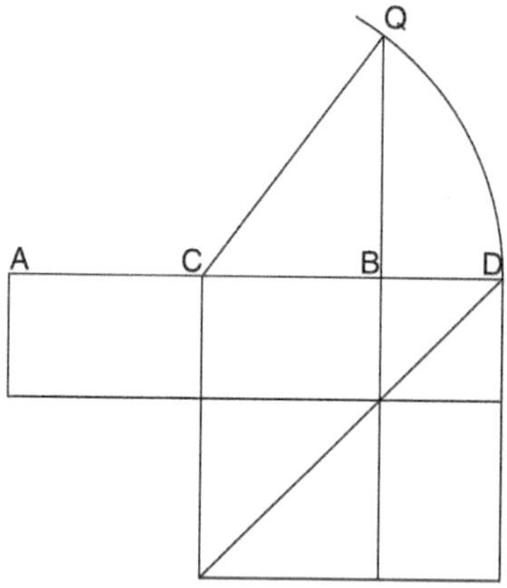

If we let the rectangle $zb \cdot bi = c^2$, $ti = x$, and $bi = a$, we have to solve geometrically the quadratic equation:

$$ax + x^2 = c^2.$$

Heath points out (*Elements*, vol. I, pp. 386–387) that the solution of a problem theoretically equivalent to the solution of a quadratic equation of this kind is presupposed in the fragment of Hippocrates' *Quadrature of lunes* (5th century B. C.) preserved in a quotation by Simplicius (fl. 500 A. D.) from Eudemus' *History of Geometry* (4th century B. C.). See Simplicius' *Comment. in Aristot. Phys.* ed. H. Diels, Berlin, 1882, pp. 61–68; see also F. RUDIO, *Der Bericht des Simplicius über die Quadratur des Antiphon und Hippokrates*, Leipzig, 1907.

Moreover as Proposition 18 is suggested by the *Elements*, II. 5, so here this problem is suggested by II. 6: *If a straight line be bisected and a straight line be added to it in a straight line, the rectangle contained by the whole with the added straight line and the added straight line together with the square on the half is equal to the square on the straight line made up of the half and the added straight line.*

Draw the straight line tkd. Since

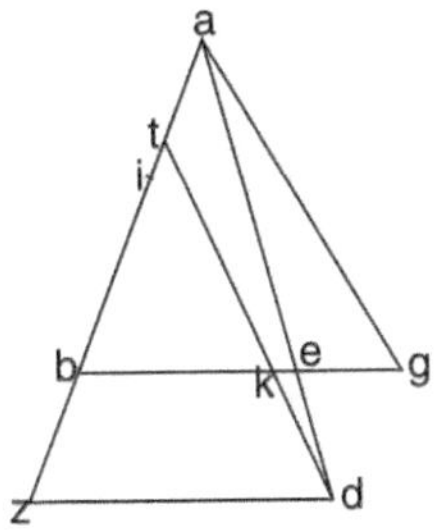

$$\text{area } zb \cdot bi = bi \cdot ti + ti^2 = \text{area } bt \cdot ti,$$

$$zb : bt = ti : ib, \qquad \text{[VII. 19]}$$

or $$zt : bt = bt : bi. \qquad \text{[V. 18]}$$

But $$zt : bt = zd : bk, \qquad \text{[VI. 4]}$$

$$\therefore zd : bk = bt : bi,$$

and $$\text{area } kb \cdot bt = \text{area } zd \cdot bi.$$

But $$\text{area } zd \cdot bi = \tfrac{1}{2} \text{ area } ab \cdot bg,$$

$$\therefore \triangle tbk = \tfrac{1}{2} \triangle abg^{88}.$$

If AB is the straight line bisected at C and BD is the straight line added, then by II. 6,

$$AD \cdot DB + CB^2 = CD^2.$$

In his solution of our problem, Robert Simson proceeds, in effect, as follows (*Elements of Euclid*, ninth ed., Edinburgh, 1793, p. 336): Draw BQ at right angles to AB and equal to c. Join CQ and describe a circle with centre C and radius CQ cutting AB produced in D. Then BD or x is found. For, by II. 6,

$$\begin{aligned} AD \cdot DB + CB^2 &= CD^2, \\ &= CQ^2, \\ &= CB^2 + BQ^2, \end{aligned}$$

$$\therefore AD \cdot DB = BQ^2,$$

whence $$(a + x)x = c^2$$

or $$ax + x^2 = c^2.$$

It was not Euclid's manner to consider more than one solution in this case.

Therefore the triangle abg is divided by a line drawn from the point d, that is, by the line tkd, into two equal parts one of which is the triangle tbk, and the other the quadrilateral $tkga$.

Q. E. F.

Leonardo now gives a numerical example. He then continues:

[If the point d were on one side, ab, produced at say, z], through z draw ze parallel to bg and meeting ag produced in e.

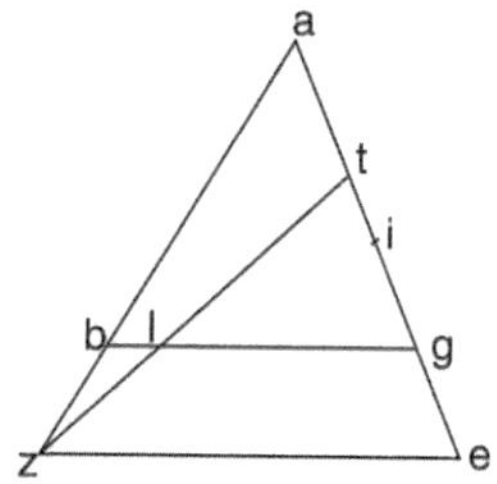

Make

$$\text{area } ze \cdot gi = \tfrac{1}{2} \text{ area } ag \cdot gb,$$

and apply a rectangle, equal to the rectangle $eg \cdot gi$, to the line gi, but exceeded by a square;

then $$eg \cdot gi = gt \cdot ti.$$

Join tz, then [this is the required line. The proof is step for step as in the first case].

Leonardo then remarks: "Que etiam demonstrentur in numeris," and proceeds to a numerical example. Thereafter he continues:

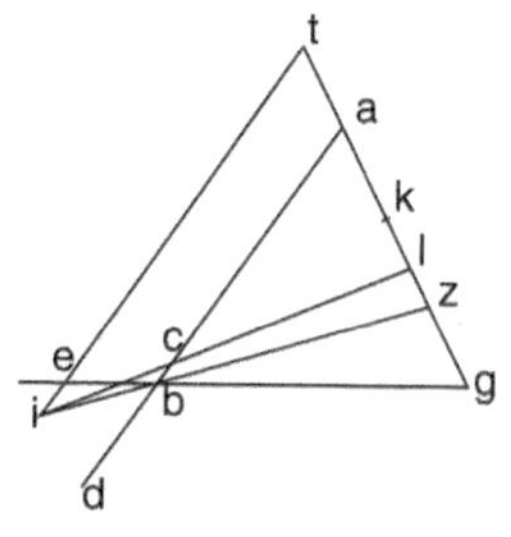

But let the sides ab, gb of the triangle be produced to d and e respectively; and let i be the given point in the angle ebd from which a line is to be drawn dividing the triangle abg into two equal parts. Join ib and produce it to meet ag in z. If $az = zg$, the triangle abg is divided into two equal parts by the line iz. But [if $az > zg$,] let za produced meet, in the point t, the line drawn through i parallel to ab.

Since $\quad za > \frac{1}{2}ag, \qquad \text{area } ab \cdot az > \frac{1}{2} \text{ area } ba \cdot ag.$

Make $\quad \text{area } it \cdot ak = \frac{1}{2}\text{area } ba \cdot ag;$

then make $\quad \text{area } al \cdot kl = \text{area } ta \cdot ak.$

Join il. Then as above the triangle abg is divided into two equal parts by the line il, one part the triangle lac, the other the quadrilateral $lcbg$.

To this statement Leonardo adds nothing further. The proof that k lies between a and z, and l between k and z, follows as in the first part.

Proposition 27.

48. *"To cut off a certain fraction of a triangle by a straight line drawn from a given point situated outside of the triangle*[117] *."* [Leonardo 11, p. 121, ll. 22–23.]

Let abg be the given triangle and d the given point outside. It is required to cut off from the triangle a certain fraction, say one-third, by a line drawn through d. Join ad, cutting bg in c. If either bc or cg be one-third of bg, then the line ad through the point d cuts off one-third of the triangle abg. But if this be not the case produce ab, ag to meet in z and e respectively the line drawn through d parallel to bg.

---

117 Some generalizations of the triangle problems in Propositions 19, 20, 26 and 27 may be remarked. Steiner, in 1827, solved the problem: *through a given point on a sphere to draw an arc of a great circle cutting two given great circles such that the intercepted area is equal to a given area.* (J. Steiner, "Verwandlung und Theilung sphärischen Figuren durch Construction," *Crelle Jl*, II (1827), pp. 56 f. *Cf.* Syllabus of Townsend's course at Dublin Univ., 1846, in *Nouvelles Annales de Mathématiques*, Sept. 1850, IX, 364; also Question 427(7) proposed by Vannson in *Nouvelles Annales...* Jan. 1858, XVII, 45; answered Aug. 1859, XVIII, 335–6.) See also Gudermann, "Über die niedere Sphärik," *Crelle Jl*, 1832, VIII, 368.

In the next year Bobillier solved, by means of planes and spheres only, the problem, *to draw through a given line a plane which shall cut off from a given cone of revolution a required volume.* (*Correspondance Math. et Physique* (Quetelet), VIe livraison, IV, 2–3, Bruxelles, 1828.)

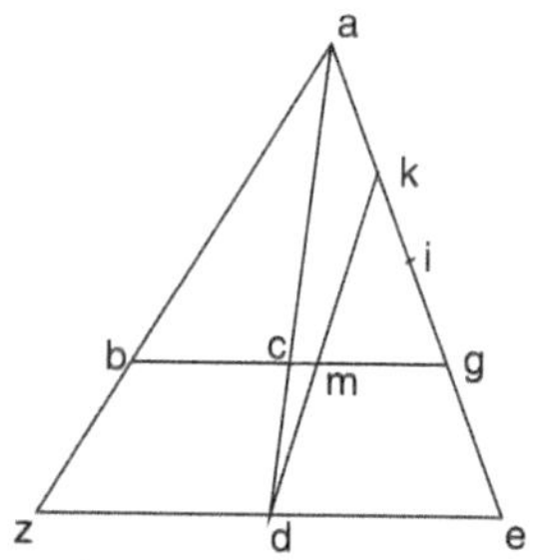

Make

$$\text{area } de \cdot gi = \tfrac{1}{3} \text{ area } ag \cdot gb,$$

and apply to the line gi a rectangle equal to the rectangle $eg \cdot gi$, but exceeded by a square; then

$$eg \cdot gi = ik \cdot kg.$$

Draw the line kmd. I say that the triangle kmg is one-third of the triangle abg.

Proof: For since

$$\text{area } eg \cdot gi = \text{ area } gk \cdot ki,$$

$$eg : gk = ki : ig.$$

Hence $$ek : gk = gk : gi. \qquad [\text{V. } 17]$$

But $$ek : kg = de : gm; \qquad [\text{VI. } 2]$$

$$\therefore ed : gm = gk : gi.$$

$$\therefore \text{ area } gk \cdot gm = \text{ area } de \cdot gi.$$

But $$\text{area } de \cdot gi = \tfrac{1}{3} \text{ area } ag \cdot gb;$$

$$\therefore \text{ area } gk \cdot gm = \tfrac{1}{3} \text{ area } ag \cdot gb.$$

And since

$$\text{area } gk \cdot gm : \text{ area } ag \cdot gb = \triangle kgm : \triangle agb^{88},$$

$$\triangle kgm = \tfrac{1}{3}\triangle agb.$$

In a similar manner any part of a triangle may be cut off by a straight line drawn from a given point, on a side of the triangle produced, or within two produced sides.

Proposition 28.

49. *"To divide into two equal parts a given figure bounded by an arc of a circle and by two straight lines which form a given angle."* [Leonardo 57, p. 148, ll. 13–14.]

"Let ABC be the given figure bounded by the arc BC and by the two lines AB, AC which form the angle BAC. It is required to draw a straight line which will divide the figure ABC into two equal parts.

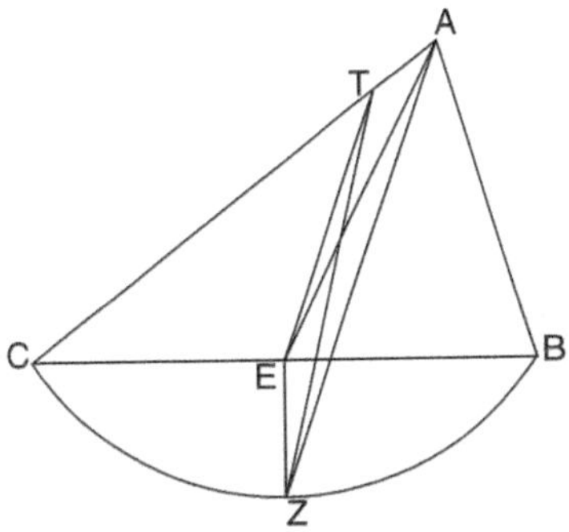

"Draw the line BC and bisect it at E. Through the point E draw a line perpendicular to BC, as EZ, and draw the line AE. Then because BE is equal to EC, the area BZE is equal to the area EZC, and the triangle ABE is equal to the triangle AEC. Then the figure $ABZE$ will equal the figure $ZCAE$. If the line AE lie in EZ produced, the figure will be divided into two equal parts $ABZE$ and $CAEZ$. But if the line AE be not in the line ZE produced, join A to Z by a straight line and through the point E draw a line, as ET, parallel to the line AZ. Finally draw the line TZ. I say, that the line TZ is that which it is required to find, and that the figure ABC is divided into two equal parts $ABZT$ and ZCT.

"For since the two triangles TZA and EZA are constructed on the same base AZ and contained between the same parallels AZ, TE: the triangle ZTA is equal to the triangle AEZ. Then, adding to each the common part AZB, we have $TZBA$ equal to $ABZE$. But this latter figure was half of the figure ABC; consequently the line ZT is the line sought and $BZCA$ is divided into two equal parts $ABZT$, TZC, which was to be demonstrated."

Leonardo's proof is practically word for word as the above. He gives two figures and in each he uses the *Greek* succession of letters.

It is doubtless to this Proposition and the next that reference is made in the account of Proclus [Art. I].

PROPOSITION **29.**

50. *"To draw in a given circle two parallel lines cutting off a certain fraction from the circle."* [Leonardo 51 (the case where the fraction is one-third), p. 146, ll. 37–38.]

"Let the certain fraction be one-third, and the circle be ABC. It is required to do that which is about to be explained.

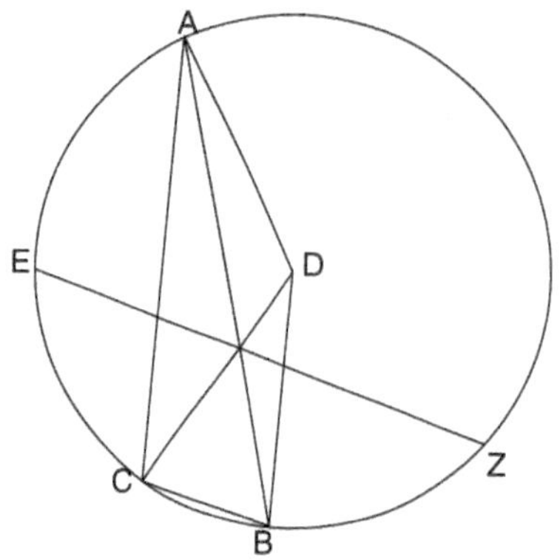

"Construct the side of the triangle (regular) inscribed in this circle. Let this be AC. Draw the two lines AD, DC and draw through the point D a line parallel to the line AC, such as DB. Draw the line CB. Divide the arc AC into two equal parts at the point E, and draw from the point E the chord EZ parallel to the line BC. Finally draw the line AB. I say that we have two parallel lines EZ, CB cutting off a third of the circle ABC, viz. the figure $ZBCE$.

"*Demonstration.* The line AC being parallel to the line DB, the triangle DAC will be equal to the triangle BAC; add to each the segment of the circle AEC; the whole figure $DAEC$ will be equal to the whole figure $BAEC$. But the figure $DAEC$ is one-third of the circle. Consequently the figure $BAEC$ is also one-third of the circle. Since EZ is parallel to CB, the arc EC will be equal to the arc BZ; but EC is equal to EA, hence EA equals ZB. Add to these equal parts the arc ECB; the whole arc AB will equal the whole arc EZ. Consequently the line AB will be equal to the line EZ, and the segment of the circle $AECB$ will be equal to the segment of the circle $ECBZ$. Taking away the common segment BC, there remains the figure $EZBC$ equal to the figure $BAEC$. But the figure $BAEC$ was one-third of the circle ABC. Then the figure $EZBC$ is one-third of the circle ABC; which was to be demonstrated.

"When it is required to cut off a quarter of a circle, or a fifth or any other definite fraction, by means of two parallel lines, we construct in this circle the side of a square or of the pentagon (regular) inscribed in the circle and we draw from the centre to the extremities of this side the two straight lines as above. (The remainder of) the construction will be analogous to that which has gone before[118]

The statement and form of discussion of this proposition are not wholly satisfactory. For "a certain fraction" in the enunciation we should rather expect "one-third," as in Leonardo; while at the conclusion of the proof might possibly occur a remark to the effect that a similar construction would apply when the certain fraction was one-quarter [by means of IV. 6], one-fifth [IV. 11], one-sixth [IV. 15], or one-fifteenth [IV. 16], but is it conceivable that Euclid added "or any other definite fraction"? Moreover the lack of definition of D and certain matters of form seem to further indicate that modification of the original has taken place in its passage through Arabian channels.

On the other hand Leonardo presents the proposition as if drawn from the pure well of Euclid undefiled. Here is his discussion. (I have substituted C for his b, and B for his g.)

"And if, by means of two parallel lines, we wish to cut off from a circle ACB, whose centre is D, a given part which is one-third, draw the line AC, the side of an equilateral triangle inscribed in the circle abg. Through the centre D draw DB parallel to this line and join CB. Bisect the arc AC at E and draw EZ parallel to bg. I say that the figure contained between the lines CB and EZ and the arcs EC and BZ is one-third part of the circle ACB.

"*Proof:* Draw the lines DA and DB and AB.

"The triangles BAC and DAC are equal. To each add the portion ABE. Then the figure bounded by the lines BA and BC and the arc AEC is equal to the sector $DAEC$ which is a third part of the circle ABC.

"Therefore the figure bounded by the lines BA and BC and the arc AEC is a third part of the circle.

"And since the lines CB and EZ are parallel, the arcs EC and BZ are equal. But arc EC is equal to arc AE. Therefore arc AE is equal to the arc BZ. To each add the arc EB, and then the arc $AECB$ will be equal to the arc $ECBZ$.

"Hence the portion $EZBC$ of the circle is equal to the portion $ABCE$. Take away the common part between the line CB and the arc BC and there remains the figure, bounded by the lines BC and EZ and the arcs CE and BZ, which is

---

[118] This problem is clearly not susceptible of solution with ruler and compasses, in such a case as when the "certain fraction," $\frac{1}{n}$, is one-seventh. In fact the only cases in which the problem is possible, for a fraction of this kind, is when n is of the form

$$2^p(2^{2^{s_1}}+1)(2^{2^{s_2}}+1)\dots(2^{2^{s_m}}+1),$$

where p, and s's (all different), are positive integers or zero, and $2^{2^{s_m}}+1 (m=1,2,\dots m)$ is a prime number. (*Cf.* C. F. Gauss, *Disquisitiones Arithmeticae*, Lipsiae, 1801, French ed., Paris, 1807, p. 489.)

the third part of the circle since it is equal to the figure bounded by the lines BA and BC and the arc AEC; quod oportebat ostendere."

In his μετρικά (III. 18) Heron of Alexandria considers the problem: *To divide the area of a circle into three equal parts by two straight lines.* He remarks that "it is clear that the problem is not rational"; nevertheless "on account of its practical use" he proceeds to give an approximate solution. By discussion similar to that above he finds the figure $BCEA$, formed by the triangle BCA and the segment CEA, to be one-third of the circle. Neglecting the smaller segment with chord BC, we have, that BA cuts off "approximately" one-third of the circle. Similarly a second chord from B might be drawn to cut off another third of the circle, and the approximate solution be completed.

PROPOSITION 30.

51. *"To divide a given triangle into two parts by a line parallel to its base, such that the ratio of one of the two parts to the other is equal to a given ratio."*

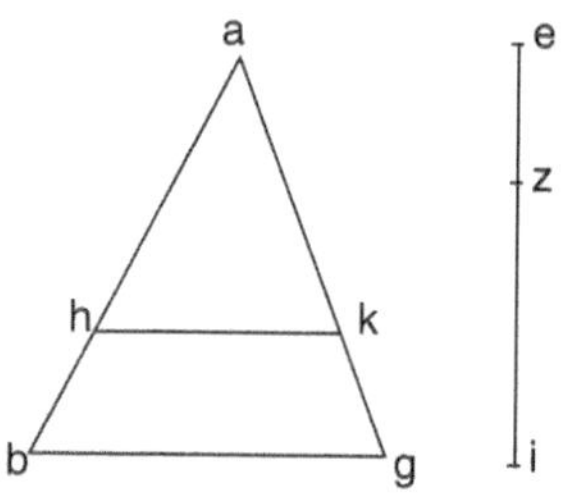

Although Leonardo does not explicitly formulate this problem or the next, the method to be employed is clearly indicated in the discussion of Proposition 5 (Art. 26).

Let abg be the triangle which is to be divided in the given ratio $ez : zi$, by a line parallel to bg. Divide ab in h such that

$$ah^2 : ab^2 = ez : ei$$[92].

Draw $hk \parallel bg$ and meeting ag in k. Then the triangles ahk and abg are similar and

$$\triangle ahk : \triangle abg = ah^2 : ab^2. \qquad \text{[VI. 19]}$$

But
$$ah^2 : ab^2 = ez : ei,$$
$$\therefore \triangle ahk : \triangle abg = ez : ei;$$
whence
$$\triangle ahk : \text{quadl.}hbgk = ez : zi; \qquad \text{[V. 16, 17]}$$

and the triangle abg has been divided as required.

PROPOSITION 31.

52. *"To divide a given triangle by lines parallel to its base into parts which have given ratios to one another."*

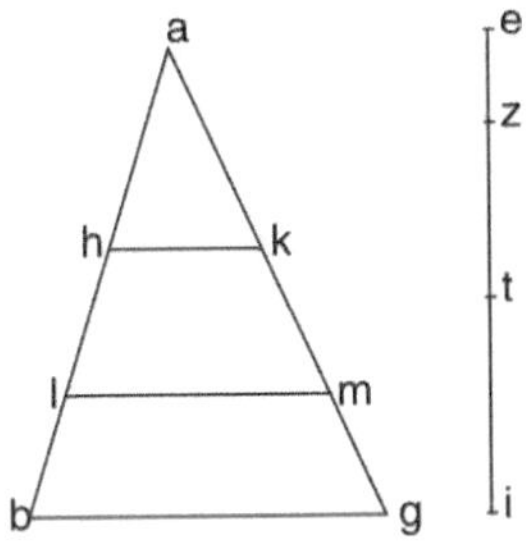

Again in the manner of Proposition 5, suppose it be required to divide the triangle abg into three parts in the ratio $ez : zt : ti$. Then determine the points h, l in ab such that

$$ah^2 : ab^2 = ez : ei;$$[92]

and $$al^2 : ab^2 = et : ei.$$

Then $$ah^2 : al^2 = ez : et.$$ [v. 16, 20]

$$\therefore \triangle ahk : \triangle alm = ez : et,$$

and $$\therefore \triangle ahk : \text{quadl.}hlmk = ez : zt.$$

Similarly

$$\triangle alm : \text{quadl.}lbgm = et : ti.$$

But $$\triangle ahk : \triangle alm = ez : et.$$

$$\therefore \triangle ahk : \text{quadl.}lbgm = ez : ti.$$ [v. 20]

Hence, $$\triangle ahk : \text{quadl.}hlmk : \text{quadl.}lbgm = ez : zt : ti,$$

and the triangle abg has been divided into three parts in a given ratio to one another. So also for any number of parts which have given ratios to one another.

PROPOSITION 32.

53. *"To divide a given trapezium by a line parallel to its base, into two parts such that the ratio of one of these parts to the other is equal to a given ratio."* [Leonardo 29, p. 131, ll. 41–42.]

Let $abgd$ be the trapezium which is to be divided in the ratio $ez : zi$ by a line parallel to the base. Produce the sides ba, gd through a and d to meet in t.

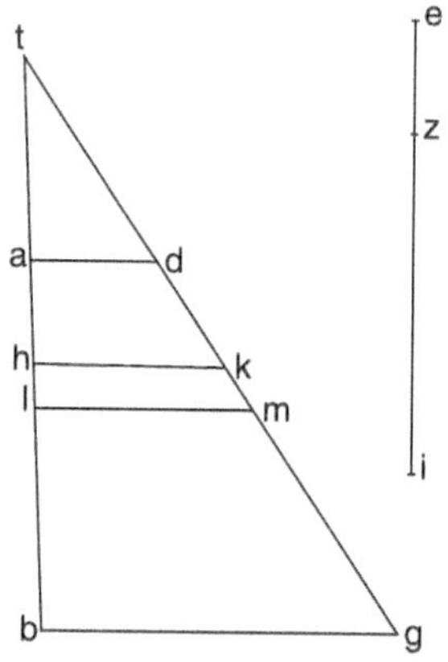

Make $$tl^2 : at^2 = zi : ez$$[92],

and $$ht^2 : (bt^2 + tl^2) = ez : ei.$$

Through l, h, draw lm, hk parallel to bg and ad. Then I say that the quadrilateral ag is divided in the given ratio, $ez : zi$, by the line hk.

Proof: For since the triangles tlm, tad are similar

$$tl^2 : at^2 = \triangle tlm : \triangle tad;$$

but $$tl^2 : at^2 = zi : ez;$$

$$\therefore zi : ez = \triangle tlm : \triangle tad.$$

Whence $$ei : ez = (\triangle tlm + \triangle tad) : \triangle tad,$$ [V. 18]

or $$ez : ei = \triangle tad : (\triangle tlm + \triangle tad).$$ [V. 16]

But by construction

$$ez : ei = ht^2 : (bt^2 + tl^2),$$

and $$ht^2 : (bt^2 + tl^2) = \triangle thk : (\triangle tbg + \triangle tlm).$$ [VI. 19]

$$\therefore ez : ei = \triangle thk : (\triangle tbg + \triangle tlm).$$

But $$\triangle thk = \triangle tad + \text{quadl.}ak.$$

Similarly

$$\triangle tbg + \triangle tlm = \text{quadl.}ag + \triangle tad + \triangle tlm.$$

$$\therefore ez : ei = (\text{quadl.}ak + \triangle tad) : (\text{quadl.}ag + \triangle tad + \triangle tlm).$$

But $$ez : ei = \triangle tad : (\triangle tad + \triangle tlm);$$

$$\therefore ez : ei = \text{quadl.}ak : \text{quadl.}ag; \qquad \text{[v. 11, 19]}$$

whence $$ez : zi = \text{quadl.}ak : \text{quadl.}hg.$$

And the trapezium has been divided in the given ratio.

Then follows a numerical example and this alternative construction and proof:

Draw mls such that,

$$ms : ls = tb^2 : ta^2,$$ [92]

and divide ml in n, such that ln is to nm in the given proportion.

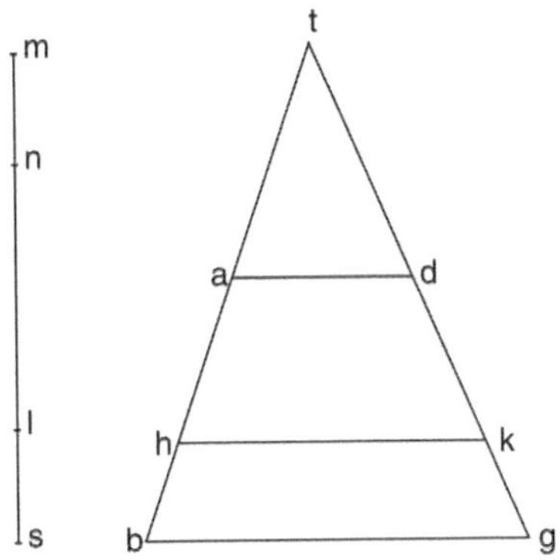

In tb determine h such that

$$th^2 : tb^2 = ns : sm.$$

Draw $hk \parallel bg$. Then,

$$\text{quadl.}ak : \text{quadl.}hg = ln : nm.$$

Proof: For, $tb^2 : ta^2 = \triangle tbg : \triangle tad;$ [VI. 1

and $ms : ls = tb^2 : ta^2.$

$\therefore ms : ls = \triangle tbg : \triangle tad.$ .... [

Again, since $tb^2 : th^2 = ms : sn,$

and $\triangle tbg : \triangle thk = tb^2 : th^2,$

$\therefore ms : ns = \triangle tbg : \triangle thk;$ .... [

$\therefore sm : nm = \triangle tbg : \text{quadl.}hg.$ [V. 16, 21] ...... [

But $sm : ls = \triangle tbg : \triangle tad$[1181],

or $ms : \triangle tbg = ls : \triangle tad,$

while $ms : \triangle tbg = ns : \triangle thk;$ [from [2

$\therefore ls : ns = \triangle tad : \triangle thk.$ .... [

From [3] $ms : \triangle tbg = nm : \text{quadl.}hg.$

But from [4] $sl : ln = \triangle tad : \text{quadl.}ak,$ [V. 16, 2

$\therefore sl : \triangle tad = ln : \text{quadl.}ak.$

But from [1] $ms : \triangle tbg = sl : \triangle tad,$

$\therefore ms : \triangle tbg = ln : \text{quadl.}ak;$

$\therefore mn : \text{quadl.}hg = ln : \text{quadl.}ak;$

$\therefore ln : nm = \text{quadl.}ak : \text{quadl.}hg.$

Hence the quadrilateral ag is divided by the line hk, parallel to the base bg, in the given proportion as the number ln is to the number nm. Which was to be done.

Then follows a numerical example.

PROPOSITION 33.

54. *"To divide a given trapezium, by lines parallel to its base, into parts which have given ratios to one another."* [Leonardo 35, p. 137, ll. 6–7.]

Let $abgd$ be the given trapezium and [let $ez : zi : it$ denote

[1181] Such mixed ratios as these (ratios of lines to areas), and others of like kind which follow in this proof, are very un-Greek in their formation. This is sufficient to stamp the second proof as of origin other than Greek. The first proof, on the other hand, is distinctly Euclidean in character.

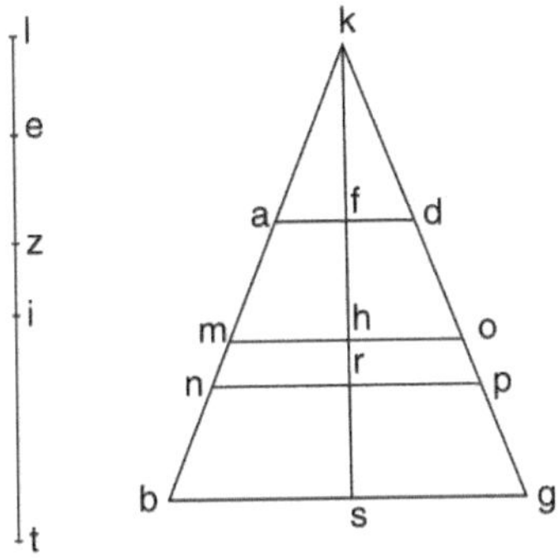

the ratios of the three parts into which the trapezium is to be divided by lines parallel to the base bg]. Let ba, gd produced meet in k and find l such that

$$bk^2 : ak^2 = tl : el.$$

Then determine m and n such that

$$bk^2 : km^2 = tl : lz,$$

and

$$bk^2 : kn^2 = tl : il.$$

Through m, n draw lines mo, np parallel to bg. In the same manner as above

$$\text{quadl.}ao : \text{quadl.}mp = ez : zi;$$

and

$$\text{quadl.}mp : \text{quadl.}ng = zi : it.$$

Then follows a numerical example in which the line $kfhrs$, perpendicular to bg, is introduced into the figure.

Here is a proof of the Proposition:

By construction, V. 16 and VI. 19,

	$\triangle kbg : \triangle kad = tl : el.$	..................[1]
So also	$\triangle kbg : \triangle kmo = tl : lz,$	..................[2]
and	$\triangle kbg : \triangle knp = tl : il$	..................[3]
From [1]	$\triangle kad : \triangle kbg = el : tl.$	
But from [2]	$\triangle kbg : \triangle kmo = tl : lz;$	
hence, by [V. 20],	$\triangle kad : \triangle kmo = el : lz,$	..................[4]
or alternately	$\triangle kmo : \triangle kad = lz : el.$	
Hence, *separando*,	quadl.$ao : \triangle kad = ez : el.$	..................[5]

So also from [2] and [3]

	$\triangle kmo : \triangle knp = lz : il;$	
and	$\triangle kmo :$ quadl.$mp = lz : iz.$	
But from [4]	$\triangle kad : \triangle kmo = el : lz;$	
therefore, by [V. 20],	$\triangle kad :$ quadl.$mp = el : iz$	..................[6]

Hence from [5], by [V. 20],

quadl.ao : quadl.$mp = ez : zi.$

Again, from [3],	quadl.$ng : \triangle kbg = ti : tl;$
and since from [1],	$\triangle kbg : \triangle kad = tl : el,$
we have	quadl.$ng : \triangle kad = it : el.$

Hence from [6], by [V. 20], we get

quadl.ng : quadl.$mp = it : zi,$

or alternately quadl.mp : quadl.$ng = zi : it.$

And since quadl.ao : quadl.$mp = ez : zi$, the trapezium ag has been divided by lines parallel to the base ag, into three parts which are in the required ratios to one another.

Q. E. F.

PROPOSITION 34.

55. "*To divide a given quadrilateral, by a line drawn from a given vertex of the quadrilateral, into two parts such that the ratio of one of these parts to the other is equal to a given ratio.*" [Leonardo 40, p. 140, ll. 36–37.]

Let $abcd$ be the given quadrilateral, and $ez : zi$ the given ratio. It is required to draw from the angle d a line to divide the quadrilateral in the ratio $ez : zi$.

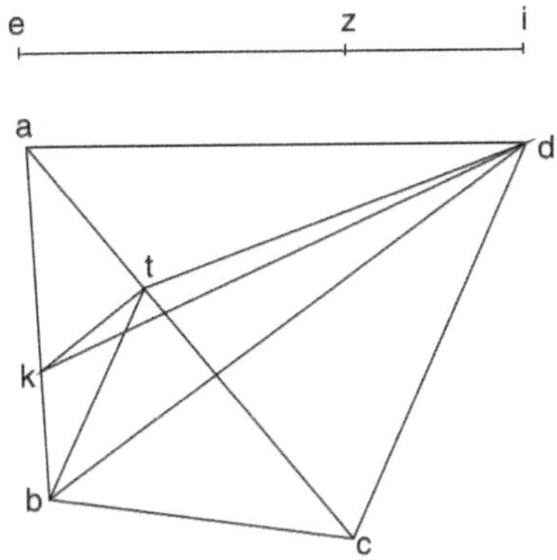

Draw the diagonal ac and on it find t such that

$$ct : at = ez : zi.$$

Draw the diagonal bd. Then if bd pass through t the quadrilateral is divided as required, in the ratio $ez : zi$.

For,

$$\triangle dct : \triangle dta = ct : ta,$$
$$= \triangle cbt : \triangle tba.$$
$$\therefore ct : ta = \triangle dcb : \triangle abd. \qquad \text{[v. 18]}$$

But $$ct : ta = ez : zi;$$
$$\therefore ez : zi = \triangle bdc : \triangle bda;$$

and the quadrilateral ac is divided, by a line drawn from a given angle, in a given ratio.

But if bd do not pass through t, it will cut ca either between c and t or between t and a. Consider first when bd cuts ct. Join bt and td. Then,

$$\text{quadl.}tbcd : \text{quadl.}tbad = ct : ta = ez : zi.$$

Draw $tk \parallel bd$ and join dk. Then

$$\text{quadl.}kbcd = \text{quadl.}tbcd;$$
$$\therefore \text{quadl.}kbcd : \triangle dak = ez : zi,$$

and the line dk has been drawn as required.

If the diagonal bd cut at, through t draw tl parallel to the diagonal bd. Join dl. Then as before,

$$ct : ta = ez : zi = \triangle dcl : \text{quadl.}abld.$$

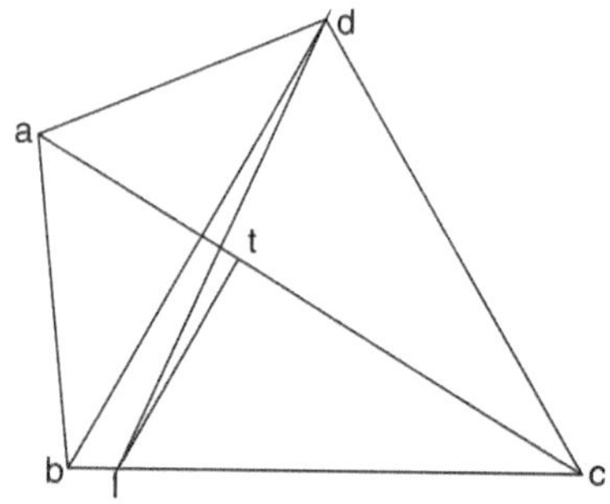

Hence in every case the quadrilateral has been divided as required by a line drawn from d. Similarly for any other vertex of the quadrilateral.

PROPOSITION 35.

56. *"To divide a given quadrilateral by lines drawn from a given vertex of the quadrilateral into parts which are in given ratios to one another."*

Although Leonardo does not explicitly formulate this problem, the method he would have followed is clear from his discussion of the last Proposition. Let $abcd$ be the quadrilateral to be divided, by lines drawn from d, into three parts in the ratios to one another of $ez : zi : it$.

Divide ca at points r, t so that

$$cr : rt : ta = ez : zi : it.$$

Through r, t draw lines parallel to bd, and meeting bc (or ab) in l and ab (or bc) in m.

Then as above dl, dm divide the quadrilateral as required.

We may proceed in a similar manner to divide the quadrilateral $abcd$, by lines drawn from the angular point d, into any number of parts in given ratios to one another.

PROPOSITION 36.

57. *"Having resolved those problems which have gone before, we are in a position to divide a given quadrilateral in a given ratio or in given ratios by a line or by lines drawn from a given point situated on one of the sides of the quadrilateral, due regard being paid to the conditions mentioned above."*

This problem, also, is not formulated by Leonardo; but from his discussion of Euclid's Propositions 16, 17 and of his own 41, the methods of construction which Euclid might have employed are clearly somewhat as follows.

Let *abcd* be the given quadrilateral and g the given point.

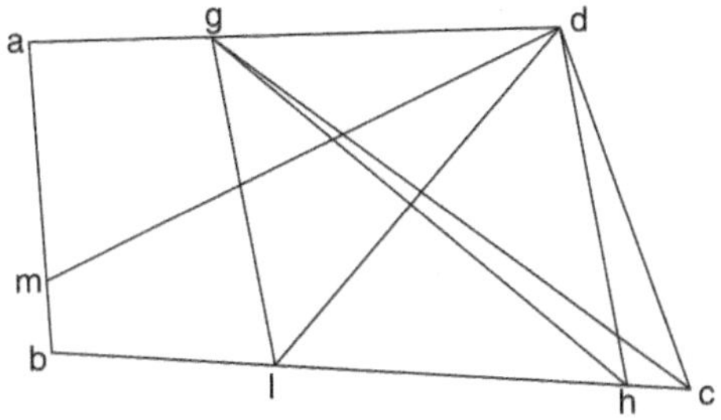

(1) Let it be required to divide *abcd* into two parts in the ratio $ez : zi$ by a line drawn through a point g in the side *ad*.

Draw *dl* such that $\triangle cld : \text{quadl.}\, lbad = ez : zi$. [Prop. 34]
Join *gl*. If $gl \parallel dc$, join *gc*, then this line divides the quadrilateral as required.

If *gl* be not parallel to *dc* draw $dh \parallel gl$, and meeting *bc* in h. Join *gh*. Then *gh* divides the quadrilateral as required.

If *dh* fell outside the quadrilateral draw $ll' \parallel cd$ (not indicated in the figure) to meet *ad* in l'. Draw $l'z' \parallel gc$ to meet *dc* in z'. Then gz' is the line required.

The above reasoning is on the assumption that *dl* meets *bc* in l. Suppose now it meet *ab* in l. Join *bd* and draw *bk* such that

$$\text{quadl.}\, bcdk : \triangle kab = ez : zi.$$

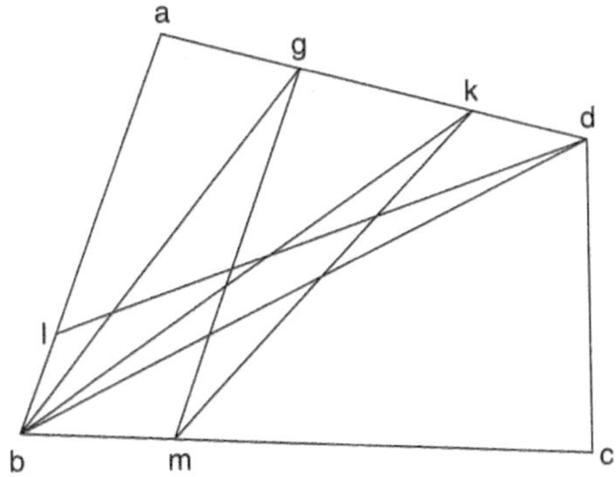

If k do not coincide with g there are two cases to consider according as k is between g and d or between g and a. Consider the former case.

Through k draw km parallel to gb and meeting bc in m. Join gm. Then this is the required line dividing the quadrilateral ac in such a way that

$$\text{quadl.}abmg : \text{quadl.}gmcd = ez : zi.$$

Similarly if k were between g and a.

(2) Let it be required to divide $abcd$ into, say, three parts in the ratios $ez : zi : it$, by lines through any point g in the side ad (first figure).

Draw dl, dm dividing the quadrilateral ac into three parts such that

$$\triangle amd : \text{quadl.}dmbl : \triangle dlc = ez : zi : it.$$

There are various cases to consider according as l and m are both on bc, both on ab, or one on ab and one on bc. The method will be obvious from working out one case, say the last.

Join gc, gl. If gl be parallel to cd, gc cuts off the triangle gdc such that

$$\triangle gdc : \text{quadl.}abcg = it : ei \qquad (= ez + zi). \qquad [\text{V. } 24]$$

If gl be not parallel to dc, draw dh parallel to gl and meeting bc in h; then gh divides the quadrilateral in such a way that

$$\text{quadl.}gdch : \text{quadl.}ghba = it : ei.$$

Then apply Proposition 34 to draw from g a line to divide the quadrilateral $abhg$ in the ratio of $ez : zi$.

Hence from g are drawn two lines which divide the quadrilateral $abcd$ into three parts whose areas are in the ratios of $ez : zi : it$.

The case when dh meets bc produced may be considered as above.

We could proceed in a similar manner if the quadrilateral $abcd$ were to be divided by lines drawn from g, into a greater number of parts in given ratios.

The enunciation of this proposition is a manifest corruption of what Euclid may have given. Such clauses as those at the beginning and end he would only have included in the discussion of the construction and proof.

After the enunciation of Proposition 36, Woepcke's translation of the Arabian MS. concludes as follows:

"End of the treatise. We have confined ourselves to giving the enunciations without the demonstrations, because the demonstrations are easy."

IV.

APPENDIX

In the earlier pages I have referred to works on Divisions of Figures written before 1500. Several of these were not published till later; for example, that of Muhammed Bagdedinus in 1570, of Leonardo Pisano in 1862 and the second edition of Luca Paciuolo's "Summa" in 1523[119]. It has been remarked that Fra Luca's treatment of the subject was based entirely upon that of Leonardo. But, on account of priority in publication, to Paciuolo undoubtedly belongs the credit of popularizing the problems on Divisions of Figures.

While few publications treat of the subject in the early part, their number increases in the latter part, of the sixteenth century. In succeeding centuries the tale of titles is enormous and no useful purpose would be served by the publication here of an even approximately complete list. It would seem, however, as if the subject matter were of sufficient interest to warrant, as completion of the history of the problems, a selection of such references in this period, (1) to standard or popular works, (2) to the writings of eminent scientists like Tartaglia, Huygens, Newton, Kepler and Euler; (3) to special articles, pamphlets or books which treat parts of the subject; (4) to discussions of division problems requiring other than Euclidean methods for their solution.

No account is taken of the extensive literature on the division of the circumference of a circle, from which corresponding divisions of its area readily flow. Considerations along this line may be found in: P. BACHMANN, *Die Lehre von der Kreisteilung und ihre Beziehungen zur Zahlentheorie*, Leipzig, 1872, 12 + 300 pp.; and in A. MITZSCHERLING, *Das Problem der Kreisteilung*, Leipzig und Berlin, 1913, 6 + 214 pp.

Except for about a dozen titles, all the books or papers mentioned have been personally examined. In many cases it will be found that only a single problem (often Euclid's Propositions 19, 20, 26 or 27) is treated in the place referred to.

Some titles in note 111 may also be regarded as forming a supplement to this list.

1539—W. SCHMID. *Das Erst* [Zweit, Dritt und Viert] *Buch der Geometria.* Nürnberg.

> "Dritter Theil, von mancherley Art der Flächen, wie dieselben gemacht und ausgetheilt werden, auch wie eine Fläche in die andern für sich selbst, oder gegen einer andern in vorgenommener Proporz, geschätzt, verändert mag werden. Theilungen und Zeichnungen von Winkeln, Figuren, ordentlichen Vielecken, die letzten, wie man leicht denken kann, nicht alle geometrisch richtig. Verwandlung von Figuren."

---

119 A synopsis of the portion of the work on divisions of figures is given on pages 106 and 275–284 of *Scritti inediti del P. D. Pietro Cossali...* pubblicati da Baldassarre Boncompagni, Roma, 1857. *Cf.* note 61.

1547—L. FERRARI. A "Cartello" which begins: "Messer N. Tartaglia, già otto giorni, cioè alli 16 di Maggio, in risposta della mia replica io riceuetti la uostra tartagliata, etc." [Milan.]

Dated June 1, 1547; a challenge to a mathematical disputation from L. F. to N. Tartaglia.

1547—N. TARTAGLIA. *Terza Risposta data da N. Tartalea... al eccellente M. H. Cardano... et al eccellente Messer L. Ferraro... con la resolutione, ouer risposta de 31 quesiti, ouer quistioni da quelli a lui proposti.* [Venice, 1547.]

Dated July 9, 1547. For the discussion between Ferrari (1522–1565) and Tartaglia (1500–1557) 6 "cartelli" by Ferrari and 6 "Risposte" by Tartaglia were published at Milan, Venice and Brescia in 1547–48[120]. They contained the problems and their solutions. These publications are of excessive rarity. Only about a dozen copies (which are in the British Museum and Italian Libraries) are known to exist. But they have been reprinted in: *I sei cartelli di matematica disfida primamente intorno alla generale risoluzione delle equazioni cubiche, di Ludovico Ferrari, coi sei contro-cartelli in risposta di Nicolo Tartaglia... comprendenti le soluzioni de' quesiti dall' una e dall' altra parte proposti... Raccolti, autografati e pubblicati da Enrico Giordani....* Milano 1876.

On pages 6–7 of the III° cartello (Giordani's edition pp. 94–95), Questions 5 and 14, proposed by Ferrari, are:—

"5. To bisect, by a straight line, an equilateral, but not equiangular, heptagon."

"14. Through a point without a triangle to draw a line which will cut off a third."

On pages 12 and 20 of the III° Risposta Tartaglia gives the solutions and assigns due credit to the treatment of problems on the Division of Figures by Luca Paciuolo. The general subject was treated much more at length by Tartaglia in a part of his "General trattato" published in 1560.

1560—N. TARTAGLIA. *La quinta parte del general trattato de' numeri et misure.* Venetia.

On folio 6 *recto* we have a section entitled "Del modo di saper dividere una figura cioè pigliar, over formar una parte di quella in forma propria." The division of figures is treated on folios 23 *verso*–44 *recto* (23–32, triangles; 32–34, parallelograms; 34–44, quadrilateral, pentagon, hexagon, heptagon, circle without the Euclid-Proclus case).

Cf. the synopsis in *Scritti inediti del P. D. Pietro Cossali chierico regolare teatino* pubblicati da Baldassarre Boncompagni, Roma, 1857, pp. 299–300.

1574—J. GUTMAN. *Feldmessung gewiss, richtig und kurz gestellt.* Heidelberg.

1574—E. REINHOLD. *Gründlicher und wahrer Bericht vom Feldmessen, samt allem, was dem anhängig, darinn alle die Irrthum, so bis daher im Messen*

[120] *Cf.* CANTOR, *Vorlesungen über Gesch. d. Math.* Bd II, 2te Aufl., 1900, pp. 490–491, where the exact dates are given.

fürgeloffen, entdeckt werden. Dessgleichen vom Markscheiden, kurzer und gründlicher Unterricht. Erffurdt.

"Der dritte Theil von Theilung der Aecker. Theilungen aller Figuren, auch des Kreises mit Exemplen und Tafeln erläutert."

1585—G. B. BENEDETTI. *Diversarum speculationum mathematicarum et physicarum liber.* Taurini.

Pages 304–307.

1604—C. CLAVIUS. *Geometria Practica.* Romae.

Pages 276–297.

1609—J. KEPLER. *Astronomia nova... commentariis de motibus stellæ Martis...* [Pragae].

"Kepler's Problem" occurs on p. 300 of this work (*Opera Kepleri* ed. Frisch, III, 401). It is: "To divide the area of a semicircle in a certain ratio by a straight line drawn through a given point on the diameter or on the diameter produced." (*Cf.* A. G. KÄSTNER, *Geschichte der Math.* IV, 256, Göttingen, 1800; M. CANTOR, *Vorlesungen* etc., II, 708, Leipzig, 1900). Kepler was led to this problem in his theory of the path of the planets. It has been attacked by many mathematicians, notably by Wallis, Hermann, Cassini, D. Gregory, T. Simpson, Clairaut, Lagrange, Bossut, and Laplace. (*Cf.* G. S. KLÜGEL, *Mathematisches Wörterbuch...* Erste Abtheilung, Dritter Theil, Leipzig, 1808; Article, "Kepler's Aufgabe." See also C. HUTTON, *Philosophical and Mathematical Dictionary.* New edition, London, 1815, 1, 703.)

1612—SYBRANDT CARDINAEL. *Hondert geometrische questien met hare solutien.* Amsterdam.

This work is also to be found at the end of JOHAN SEMS ende IAN DOU *Practijck des landmetens.* Amsterdam, 1616. Another edition: Tractatus geometricus. Darinnen hundert schöne ... Questien [übersetzt] durch Sebastianum Curtium. Amsterdam, 1617; Questions 78, 90–93.

With these problems Huygens (1629–1695) busied himself when about 17 or 18 years of age. *Cf. Oeuvres complètes de Chr. Huygens*, Amsterdam, XI, 24 and 29, 1908.

I have elsewhere (*Nieuw Archief*, 1914) shown that Sybrandt Cardinael's work was translated into English, rearranged and published as an original work by Thomas Rudd (1584?-1656): *A hundred geometrical Questions with their solutions and demonstrations.* London, M.DC.L.

1615—LUDOLPH VAN CEULEN. *Fundamenta arithmetica et geometrica cum eorundem usu ... e vernaculo in Latinum translata a W. S[nellio], R. F.* Lugduni Batavorum.

Contains several problems on Change, and Division, of Figures.

1616—J. SPEIDELL. *A geometricall Extraction or a compendiovs collection of the chiefe and choyse Problemes, collected out of the best, and latest writers. Wherevnto is added about 30 Problemes of the Authors Invention, being for the most part, performed by a better and briefer way, than by any former writer.* London.

Another edition, 1617; second edition "corrected and enlarged," London, 1657; "Now followeth [pp. 84–125] a compleat Instruction of the diuision of all right lined figures.... Very pleasant and full of delight in practise: Also, most profitable to all surveighers, or others that are desirous to make any Inclosure."

1619—A. ANDERSON. *Exercitationum mathematicarum Decas prima. Continens, Questionum aliquot, quae nobilissimorum tum hujus tum veteris aevi, Mathematicorum ingenia exercuere, Enodationem.* Parisiis.

Problems in division of a triangle, with reference to Clavius (1604). *Cf. The Ladies' Diary*, London, 1840, pp. 55–56.

1645—C. HUYGENS. *Oeuvres Complètes*, XI, 1908, pp. 26–27; 219–225.

Solution of "Datum triangulum, ex puncto in latere dato, bifariam secare" (1645); two solutions of "Triang. ABC, sectus utcumque lineâ DE, dividendus est aliâ lineâ, FG, ita ut utraque pars DBE et ADEC bifariam dividatur" (1650–1668). See also note under 1687—J. Bernoulli.

1657—F. VAN SCHOOTEN. *Exercitationvm mathematicarum liber primus continens propositionum arithmeticarvm et geometricarvm centuriam.* Lugd. Batav.

Prop. L, pp. 107–110. Dutch edition, Amsterdam, 1659. pp. 107–110. Concerning a Schooten MS. of 1645–6, used by Huygens and of interest in this connection, *cf.* C. HUYGENS, *Oeuvres Complètes*, tome XI, 1908, p. 13 ff.

1667—D. SCHWENTER. *Geometriae Practicae novae et auctae Libri IV ... mit vielen nutzlichen Additionen und neuen Figuren vermehret durch G. A. Böcklern.* Nürnberg.

"Von Austheilung der Figuren in gleiche und ungleiche Theil," pp. 269–279; p. 350; the problem on this last page is taken from B. BRAMER, *Trigonometria planarum mechanica*, Marpurg, 1617, p. 99. "Von Austheilungen der Aecker Wiesen,... " pp. 567–583.

1674—C. F. M. DESCHALES. *Cursus seu mundus mathematicus.* Lugduni.

"De figurarum planarum divisione," I, 371–381; second edition, 1690.

1676—I. NEWTON. *Arithmetica Universalis.* Cantabrigiae, MDCCVII.

Prob. X, p. 126 (Prob. XX, pp. 254–255 of the 1769 edition). This problem was discussed in a lecture delivered October, 1676 (see *Correspondence of Sir Isaac Newton and Professor Cotes...* by J. Edleston, London, 1850, p. xciii).

1684—T. STRODE. *A Discourse of Combinations, Alternations and Aliquod Parts* by John Wallis. London, 1685.

On pages 163–164 is printed a letter, dated Nov. 1684, from Strode to Wallis. It discusses two problems on divisions of a triangle.

1687—J. BERNOULLI. "Solutio algebraica problematis de quadrisectione trianguli scaleni, per duas normales rectas." *Acta Eruditorum*, 1687, pp. 617–623.

Also in *Opera*, Genevae, 1744, I, 328–335; see further II, 671. In the solution of this question Bernoulli is led to the intersection of a conic and a curve of the fourth degree, that is, to an equation of the eighth degree. And yet, in the seventh edition of ROUCHÉ et COMBEROUSSE, *Traité de Géométrie*, Paris, 1900, we find Problem 453 is: "Partager un triangle quelconque en quatre parties équivalentes par deux droites perpendiculaires entre elles!" The problem was solved by L'Hospital before 1704, the year of his death, in a posthumous work, *Traité analytique des Sections coniques*, Paris, 1707, pp. 400–407. As the result of correspondence in *L'Intermédiaire des Math.*, tomes I–VII, 1894–1900, Questions 3 and 587, Loria wrote the history of the problem: "Osservazioni sopra la storia di un problema pseudo-elementare." *Bibl. Math.*, 1903 (3), IV, 48–51. Leibnitz's name appears in this connection. See note on 1645—C. Huygens.

1688—J. OZANAM. *L'usage du compas de proportion expliqué et demontré d'une manière courte et facile, et augmenté d'un Traité de la division des champs.* Paris.

"Division des champs," pp. 89–138. Ouvrage revu, corrigé et entièrement refondu par J. G. Garnier. Paris, 1794, pp. 165–257.

1694—S. LE CLERC. *Traité de Géométrie sur le terrain* at end of *Géométrie pratique, ou pratique de la géométrie sur le papier et sur le terrain.* Amsterdam.

1699—J. OZANAM. *Cours de mathématique*, nouv. éd. tome 3. Paris.

Pages 23–64. German translation: *Anweisung, die geradlinichten Figuren nach einen gegebenen Verhältniss ohne Rechnung zu theilen.* Frankfurt u. Leipzig, 1776.

1704—GUISNÉE. *Application de l'algèbre à la géométrie.* Paris.

Although the "approbation" signed by Fontenelle is dated "15 Juillet 1704" the work was first published in 1710; second edition "revûe, corrigée et considérablement augmentée par l'auteur," Paris, 1733, pp. 42–47; analytic discussion only.

1739—*l'abbé* DEIDIER. *La science des géométres* (*sic*) *ou la théorie et la pratique de la géométrie.* Paris.

"De la géodésie ou division des champs," pp. 279–320; divisions of triangles, rectangles, trapeziums, polygons.

1740—N. SAUNDERSON. *Elements of Algebra in ten books*, vol. 2. Cambridge.

Pages 546–554.

1747—T. SIMPSON. *Elements of Plane Geometry.* London.

Pages 151–152; new ed., London, 1821, pp. 207–208; taken from Newton (1676).

1748—L. EULER. *Introductio in analysin infinitorum.* Tomus secundus. Lausanne.

Chapter 22: "Solutio nonnullorum problematum ad circulum pertinentium." Three of the eight problems which Euler here discusses by the method of trial and error, and tables of circular arcs and logarithmic sines and tangents, are of interest to us. These are: Problem 2, "To find the sector of the circle ACB which is divided by the chord AB into two equal parts, so that the triangle ACB shall be equal to the segment AEB." Problem 4, "Given the semi-circle $AEDB$, to draw from the point A a chord AD which will divide the semi-circle into two equal parts." Problem 5, "From a point A of the circumference of a circle, to draw two chords AB, AC which shall divide the area of the circle into three equal parts." (Heron, *cf.* Art. 50.) Gregory (1840) considers these problems at the close (pp. 186–188) of his Appendix.

For other editions of Euler's "Introductio," tomus 2, see *Verzeichnis der Schriften Leonhard Eulers.* Bearbeitet von G. Eneström. Erste Lieferung, Leipzig, 1910.

1752—T. SIMPSON. *Select Exercises for young proficients in the mathematics.* London.

Problem XLII, pp. 145–6; new ed. by J. H. Hearding. London, 1810, pp. 148–9.

1754—J. LE R. D'ALEMBERT. *Encyclopédie ou Dictionnaire raisonné des sciences... mis en ordre et publié par M. Diderot...; et quant à la partie mathématique par M. d'Alembert.* Paris.

Article "Géodésie"; mostly descriptive of methods of Guisnée (1704) and Clerc (1694).

1768—J. A. EULER. "Auflösung einiger geometrischen Aufgaben," *Abhandlungen der Churfürstlich-baierischen Akademie der Wissenschaften*, v, 165–196.

Erste Aufgabe, pp. 167–182: "Man soll zeigen, wie eine jede geradlinichte Figur durch Parallellinien in eine gegebene Anzahl gleicher Theile zerschnitten werden kann." Zweite Aufgabe, pp. 182–187: "Eine Zirkel-fläche durch parallellinien in eine gegebene Anzahl gleicher Theile zu zerschneiden." Dritte Aufgabe, pp. 187–196: "Die Höhe und Grundlinie einer aufrecht-stehenden geschlossenen Parabelfläche ist gegeben, man soll dieselbe durch Parallellinien in n gleiche Theile zerschneiden." Discussion mostly analytic.

1772(?)—*J. H. Lamberts deutscher gelehrter Briefwechsel.* Herausgegeben von Joh. Bernoulli. Band 2, Berlin, 1782.

Pages 412–13, undated fragment of a letter from Lambert to J. E. Silberschlag. Analytic solution by quadratic equation of the problem: "Ein Feld *ABCD* welches in *ABFE* Wiesen, in *EFCD* Ackerfeld ist, soll durch eine gerade Linie *KM* so getheilt werden, dass so wohl die Wiesen als das Ackerfeld in beliebiger Verhältniss getheilt werde." [*ABCD* is a quadrilateral and *EF* is a straight line segment joining points on the opposite sides *AD*, *BC* respectively.]

In the *Journal of the Indian Mathematical Society*, 1914, VI, 159, N. P. Pandya proposed as Question 563: "Given two quadrilaterals in the plane of the paper show how to draw a straight line bisecting them both." A solution by means of common tangents to hyperbolas was offered in 1915, VII, 176.

1783—J. T. MAYER. *Gründlicher und ausführlicher Unterricht zur praktischen Geometrie,* 3. Teil. Göttingen.

Pages 215–303: "Theilung der Felder durch Rechnung, Theilung der Felder durch blose Zeichnung, Anwendung der Theilungsmethoden auf mancherley, in gemeinen Leben vorkommende Fälle"; dritte Auflage, 1804, pp. 232–337.

1793—J. W. CHRISTIANI. *Die Lehre von der geometrischen und ökonomischen Vertheilung der Felder, nach der Dänischen Schrift des N. Morville bearbeitet von J. W. Christiani.* Preface by A. G. Kästner. Göttingen.

1795—*Gentleman's Diary*, London.

No. 54, 1794, p. 47, Question 691 by J. Rodham: "Within a given triangle to find a point thus, that if lines be drawn from it to cut each side at right angles, the three parts into which the triangle thus becomes divided, shall obtain a given ratio." Solution by hyperbolas in No. 55, 1795, pp. 37–38. See also Davis's edition of the *Gentleman's Diary*, vol. 3, London, 1814, pp. 233–4.

1801—L. PUISSANT. *Recueil de diverses propositions de géométrie résolues ou démontrées par l'analyse algébrique suivant les principes de Monge et de Lacroix.* Paris.

Pages 33–36; German ed., Berlin, 1806; second French ed., Paris, 1809, pp. 107–111; third ed., Paris, 1824, pp. 139–142.

1805—M. HIRSCH. *Sammlung geometrischer Aufgaben*, Erster Theil. Berlin.

"Theilung der Figuren durch Zeichnung," pp. 14–25; "Theilung der Figuren durch Rechnung," pp. 42–53; Reprint, 1855; English edition translated by J. A. Ross and edited by J. M. F. Wright. London, 1827.

1807—A. BRATT. *Problema geometricum triangulum datum a dato puncto in 2 partes aequales secandi.* Greifswald.

This title is taken from C. G. KAYSER, *Bücher-Lexicon*, Erster Teil, Leipzig, 1834.

1807—J. P. CARLMARK. *Triangulus datus a dato puncto in 2 partes aequales secandus.* Greifswald.

This title and the next two are taken from E. WÖLFFING, *Math. Bücherschatz*, 1903.

1809—J. KULLBERG. *Problema geometricum triangulum datum e quovis dato puncto in 2 partes aequales secandi.* Diss. Lund.

1810—J. KULLBERG. *Problema geometricum triangulum quodcunque datum in 2 aequales divisum iterum in partes aequales ita secandi, ut rectae secantes angulum constituant rectum.* Diss. Upsala.

1811—J. P. GRÜSON. *Geodäsie oder vollständige Anleitung zur geometrischen und ökonomischen Feldertheilung.* Halle.

1819—L. BLEIBTREU. *Theilungslehre oder ausführliche Anleitung, jede Grundfläche auf die zweckmässigste Art ... geometrisch zu theilen.* Frankfurt am Main.

1821—J. LESLIE. *Geometrical Analysis and Geometry of Curve Lines* Edinburgh.

Pages 64–66.

1823—A. K. P. VON FORSTNER. *Sammlung systematisch geordneter und synthetisch aufgelöster geometrischer Aufgaben.* Berlin.

"Theilung der Flächen, mittelst der Proportion und der Aehnlichkeit," pp. 310–371.

1827—*Correspondance mathématique et physique* publié par A. Quetelet, tome III.

Page 180: "On donne dans un plan un angle et un point, et l'on demande de faire passer par le point une droite qui coupe les cotés de l'angle, de manière que l'aire interceptée soit de grandeur donnée." Solution by Verhulst, pp. 269–270. Answer by Bobillier, tome IV, pp. 2–3. Generalizing his solution, he gets the result: "tous les plans tangens d'un hyperboloïde à deux nappes, interceptent sur le cône asymptotique des volumes équivalens." Compare note 117.

1831—P. L. M. BOURDON. *Application de l'algèbre à la géométrie comprenant la géométrie analytique à deux et à trois dimensions*, troisième édition. Paris.

Pages 46–54; 5e éd., Paris, 1854, pp. 33–41; 8e éd. rev. par Darboux, Paris, 1875, pp. 30–38. Analytic discussion only.

1831—H. v. HOLLEBEN, und P. GERWIEN. *Geometrische Analysis.* Berlin, 2 Bde, 1831–1832.

"Theilungen," I, 184–191; II, 144–151.

1837—G. RITT. *Problèmes d'applications de l'algèbre à la géométrie avec les solutions développées*, 2e partie. Paris.

Pages 108–109.

1840—O. GREGORY. *Hints theoretical, elucidatory and practical, for the use of Teachers of elementary Mathematics and of self-taught students; with especial reference to the first volume of Hutton's course and Simson's Euclid, as Text-Books. Also a selection of miscellaneous tables, and an Appendix on the geometrical division of plane surfaces.* London.

"Appendix: Problems relative to the division of Fields and other surfaces," pp. 158–188; partly taken from Hirsch (1805). See also Euler (1748).

1844—DRESER. *Die Teilung der Figuren.* Darmstadt.

This title is taken from E. WÖLFFING, *Math. Bücherschatz*, 1903.

1847—R. POTTS. *An appendix to the larger edition of Euclid's Elements of Geometry; containing... Hints for the solution of the Problems...* Cambridge and London.

Ex. 91, pp. 72–73.

1852—H. CH. DE LA FRÉMOIRE. *Théorèmes et Problèmes de Géométrie élémentaire*, seconde éd. revue et corrigée par E. Catalan. Paris.

Pages 107–108; 6e éd. par Catalan, Paris, 1879, pp. 190–191.

1852—F. RUMMER. *Die Verwandlung und Theilung der Flächen in einer Reihe von Constructions- u. Berechnungs-Aufgaben.* Mit 3 Steintafeln. Heidelberg. 6 + 90 pp.

1855—P. KELLAND. "On Superposition." *Transactions of the Royal Society of Edinburgh*, 1885, XXI, 271–273 + 1 pl.

This paper deals, for the most part, with solutions of the following problem proposed to Professor Kelland by Sir John Robison: "From a given square one quarter is cut off, to divide the remaining gnomon into four such parts that they shall be capable of forming a square." In the *Transactions*, 1891, XXXVI, 91–95, + 2 pls., Robert Brodie has a paper entitled "Professor Kelland's Problem on Superposition."

1857—E. CATALAN. *Manuel des Candidats à l'école polytechnique.* Paris, Tome I.

Pages 233–4: "To divide a circle into two equal parts by means of an arc with its centre, A, on the circumference of the given circles." This is stated by A. REBIÈRE (*Mathématique et Mathématiciens*, 2e éd., Paris, 1893, p. 519) under the form: "Quelle doit être la longueur de la longe d'un cheval pour qu'en la fixant au contour d'un pré circulaire l'animal ne puisse tondre que la moitié du pré?"

The solution of this problem leads to a transcendental equation

$$\sin x - x\cos x = \frac{\pi}{2},$$

where x is the angle under which the points of section of the circumferences are seen from A. Catalan finds $x = 109°11'18''$, correct to within a second of arc.

Cf. *L'Intermédiaire des Mathématiciens*, 1914, Question 4327, XXI, 5, 69, 90, 115, 180.

1863—J. McDowell. *Exercises on Euclid and in Modern Geometry.* Cambridge.

No. 157, pp. 145–6; 3rd ed. 1881, p. 118.

1864—*Educational Times Reprint*, Vols. 1, 40, 44, 66, 68, 69; new series, Vol. 1; 1864–1910.

The problems here solved are Euclid's 19, 20, 26, 27: No. 1457 (I, 49, old edition, 1864) proposed by R. Palmer; solution by Rutherford who states that it was also published in Thomas Bradley's *Elements of Geometrical Drawing*, 1861—Nos. 7336 and 7369 (XL, 39, 1884) proposed by W. H. Blythe and A. H. Curtis; solutions by G. Heppel and Matz—No. 8272 (XLIV, 92, 1886) proposed by E. Perrin; solution by D. Biddle—No. 12973 (LXVI, 29, 1897) proposed by Radhakrishnan; solution by I. Arnold—No. 13460 (LXVIII, 35, 1898) proposed by I. Arnold; solution by W. S. Cooney, etc.—No. 13647 (LXIX, 42, 1898) proposed by I. Arnold; solution by W. C. Stanham—No. 16747 (new series XVIII, 46, 1910) proposed by I. Arnold; solution by proposer, by Euclid's *Elements* Bk I.

1864—H. Hölscher. *Anleitung zur Berechnung und Teilung der Polygone bei rechtwinkligen Koordinaten.* Berlin and Charlottenburg.

This work and the two following are representative of those which treat of Divisions of Figures by computation, rather than by graphical methods: (1) F. G. Gauss, *Die Teilung der Grundstücke, insbesondere unter Zugrundelegung Koordinaten*, 2 Auflage, Berlin, 1890; (2) L. Zimmerman, *Tafeln für die Teilung der Dreiecke, Vierecke, und Polygone*, Zweite vermehrte und verbesserte Auflage, Liebenwerda, 1896; 118+64 pp.

1870—F. Lindman. "Problema geometricum." *Archiv der Math. u. Phys.* (Grunert), Bd 51, 1870, pp. 247–252.

1879—P. M. H. Laurent. *Traité d'algèbre à l'usage des candidats....* Troisième édition. Paris.

Tome I, p. 191: "To divide a triangle into two equal parts by the shortest possible line." Solutions in *L'Intermédiaire des Mathématiciens*, 1902, IX, 194–5. See also F. G. M., *Exercices de Géométrie*, Cinquième édition. Tours et Paris, 1912, p. 802.

1892—H. S. Hall and F. H. Stevens. *Key to the Exercises and Examples contained in a Text-Book of Euclid's Elements.* London.

Ex. 7, 8, 10, 11, pp. 163–164.

1894—G. E. CRAWFORD. "Geometrical Problem." *Proc. Edinb. Math. Soc.*, Vol. 13, 1895, p. 36.

Paper read Dec. 14, 1894.

1899—W. J. DILWORTH. *A New Sequel to Euclid.* London.

Ex. XXXV, p. 190.

1901—A. LARMOR. *Geometrical Exercises from Nixon's 'Euclid Revised.'* Oxford.

Ex. 15, p. 122.

1902—C. SMITH. *Solution of the Problems and Theorems in Smith and Bryant's Elements of Geometry.* London.

Ex. 121, pp. 177–178; T. Simpson's solution and another.

1910—H. FLÜKIGER. *Die Flächenteilung des Dreiecks mit Hilfe der Hyperbel.* Diss. Bern. 50 pp. + 3 plates.

1910—R. ZDENEK. "Halbierung der Dreiecksfläche." Wien, *Zeitschrift für das Realwesen*, Jahrgang XXXV, Heft 10, 8 pp.

Discussion by projective geometry leading to hyperbolic arcs.

1911—D. BIDDLE. Problem 17197, *Educational Times*, London, November, LXIV, 475.

"Divide a square into five right-angled triangles, the areas of which shall be in arithmetic progression." Solutions in the Educational Times Reprint, new series, XXVI, III, 1914.

INDEX OF NAMES

In the following list, references are given to paragraph and footnote (= n.) numbers, except in the case of the Appendix (= App.) where the numbers are the years in the chronological list. App., without a number, refers to the introductory paragraphs on page 79.

[1]All the "Muhammeds" of Bagdad referred to in this volume are here supposed to be indicated by this single name.